D1357195

A Guide to Analog ASICs

A Guide to Analog ASICs

Paul M. Brown, Jr.

ACADEMIC PRESS, INC.
Harcourt Brace Jovanovich, Publishers
San Diego New York Boston
London Sydney Tokyo Toronto

Front cover photograph: Complete standard cell array. Courtesy of CLASIC–National Semiconductor.

All circuits and techniques presented in this book are intended solely for the illustration of concepts. The author assumes no responsibility for the use of circuitry or its suitability for a particular application. No license under any patent or other right is conveyed and no representation is made that the use of any circuitry or technology presented in this book is free from patent infringement.

It is the total responsibility of the reader to obtain any necessary licenses to use any circuitry presented herein and to test and/or modify the circuitry as necessary to insure its suitability for a particular application.

The author has made a diligent effort to insure the correctness of the material presented in this book. However, the author assumes no liability for errors or omissions.

This book is printed on acid-free paper. ♾

Academic Press, Inc.
San Diego, California 92101

United Kingdom Edition published by
Academic Press Limited
24–28 Oval Road, London NW1 7DX

Library of Congress Cataloging-in-Publication Data

Brown, Paul M., Jr.
A guide to analog ASICs / Paul M. Brown, Jr.
p. cm.
Includes index.
ISBN 0-12-136970-6
1. Application specific integrated circuits. 2. Linear integrated circuits--Design and construction--Data processing. 3. SPICE (Computer program) I. Title.
TK7874.6.B76 1991
621.381'5--dc20 91-4315
CIP

PRINTED IN THE UNITED STATES OF AMERICA
91 92 93 94 9 8 7 6 5 4 3 2 1

Contents

Preface

Application specific integrated circuits (ASICs), both analog and digital, have become standard system-level building blocks. ASIC vendors have attempted to provide tools that they hope will enable relatively inexperienced IC designers (i.e., systems engineers) to design sophisticated custom integrated circuits. This philosophy has been more successful in digital technology than in analog. Significantly more art is involved with analog design and far fewer computerized tools are available. Almost every analog ASIC vendor offers different semiconductor technologies, tool sets, documentation (usually lacking in detail and not providing the proper background and guidelines), and varying levels of engineering support. The result is that many engineers who could use analog ASICs lack the technical information to do so. They are not sure when custom analog ICs are cost effective or which vendor will best serve their needs. In addition, many engineers do not have adequate analog design experience, especially with integrated circuits. Consequently, many who could benefit from analog ASIC technology do not use it while others have bad experiences that could have easily been avoided.

Several books cover the subject of analog IC design in general, but none of them specifically address analog custom/semicustom design and the limitations imposed by that technology. This book is intended to be a working reference for any engineer or engineering manager who regularly uses analog custom technology or is contemplating using it in a product for the first time. A detailed analysis of the technology, vendor selection process, and cost trade-offs involved in deciding whether or not to use a custom analog product is presented. Other important considerations, such as whether to design the circuit in-house, use a design center, or use the vendor's design resources, are discussed. Key technical subjects relating to IC design are developed, where appropriate, and keyed to well-known definitive works to present a fresh overview of pertinent material without merely rehashing this material in unnecessary detail.

Basic analog design concepts are presented with a threefold purpose: first, to develop the reader's understanding of analog design techniques; second, to help the reader understand the capabilities and limitations of analog custom/semicustom technology; and third, to help the reader select the optimum technology and product to use for the particular purpose at hand. In addition, this book should serve as a general analog design reference.

The subject of analog design is relatively complex. It is my intent to reduce these complexities to a meaningful and easy to understand set of concepts while maintaining a sufficient level of rigor to insure the accuracy of these concepts. A heavy emphasis is placed on modeling and computer simulation. Nearly everyone engaged in serious design work has access to SPICE. All analog ASIC vendors provide some level of support for this universal analog design tool. SPICE is so useful and pervasive that it has all but replaced the time honored tradition of breadboarding analog circuitry. The use of SPICE in analog design is very important and is the focus of a great deal of this book.

Chapter 1 provides a historical look at the evolution of analog ASICs. Major analog ASIC vendors and their products are introduced. Typical development cycles and the cost effectiveness of the various types of ASICs are discussed.

Chapter 2 provides a global overview of basic semiconductor technology and IC fabrication techniques focusing on bipolar technology. The capabilities and limitations of linear IC design are reviewed and related to process constraints.

Chapter 3 presents a detailed look at the components available on the chip. The inherent limitations and advantages of integrated resistors, capacitors, transistors, diodes, and metal interconnections are discussed.

Chapter 4 presents a detailed introduction to SPICE and several *npn* and *pnp* transistor models. The transistor models and SPICE analysis will be used in Chapter 5.

Chapter 5 illustrates circuit design concepts using the integrated components described in Chapter 3 and SPICE analysis techniques presented in Chapter 4. First, the differences in economics and design philosophy between discrete and integrated design are presented and related to the components available in each medium. These points are then related to considerations used in partitioning a system and defining the custom integrated circuit. Next, integrated components are used to construct the basic circuit building blocks, such as current sources, voltage references, amplifiers, comparators, oscillators, timers, and nonlinear circuits. Their performance is verified using SPICE, and simulation examples are given.

Chapter 6 discusses IC layout considerations and their potential impact on circuit performance.

Chapter 7 presents packaging options and some of the issues that must be considered in selecting ASIC packaging. Points such as alternate sourcing, thermal performance, testability, high-frequency performance, mechanical considerations, size, and cost are discussed.

Chapter 8 discusses electrical and environmental testing of the custom IC to guarantee that it will perform as desired in its system role.

Chapter 9 discusses the economics of analog ASICs and guidelines for partitioning a system and defining a custom IC are given.

Chapter 10 presents guidelines for selecting a custom IC vendor.

I would like to thank Kent Albertson and Diane Delute for the time they spent reviewing the manuscript and their many valuable comments. Love to my wife Carolyn, daughter Kendra, and sons Paul and Robert for their patience and understanding while I worked on this project. Special thanks to Paul for the countless hours he kept me company by creating art and paper airplanes from the discarded drafts of the manuscript.

Paul M. Brown, Jr.

CHAPTER 1

An Introduction to Analog ASICs

1.0 A Historical Look at Analog ASICs

At the dawn of the electronic age, all electronic components were made separately with different techniques and technologies. The methodology used to manufacture a particular type of component was selected to optimize the manufacturing cost and electrical characteristics of that component. The components were subsequently purchased by the end user and wired together into a circuit using pliers, wire, soldering irons, and terminal strips. Not only did all circuits have to be designed component by component, but a great deal of skilled labor was required to assemble electronic equipment. The complexity, size, and capability of this equipment were severely limited. The cost was also very high for even the simplest apparatus such as a broadcast receiver or a black-and-white television.

Printed circuit boards (PC boards), introduced in the late 1940s, were the first major step toward the integration of circuitry. This technology replaced point-to-point wiring and terminal strips with a phenolic or fiberglass board with etched copper lands that provided the circuit interconnections. This technology provided ruggedized component mounting, simultaneous interconnection of components with wave soldering techniques, and greatly facilitated mass production.

Printed circuit boards, along with the development of semiconductors, opened the door to the miniaturization of electronic circuitry.

The demonstration of the first monolithic integrated circuit in 1958 paved the way for the simultaneous manufacture and interconnection of electronic components by diffusing impurities into a silicon wafer. Transistors, resistors, and small capacitors could now be manufactured concurrently using the same manufacturing process. This process, however, was designed to optimize the performance of the *npn* transistor. Components such as *pnp* transistors, field-effect transistors, resistors, and capacitors had serious performance limitations. Other components such as large capacitors, transformers, and inductors were not practical to integrate. The various components manufactured by this diffusion process were finally interconnected with a metal pattern analogous to that used on a PC board. Thus, an entire printed circuit board can now be fabricated at once on a tiny square or "chip" of silicon.

Standard-product integrated circuits were developed to perform commonly used functions such as logic gates and amplifiers. These integrated circuits replaced large amounts of discrete circuitry with components the size of a single transistor. They allowed a great deal more circuitry to be placed on a given-size PC board, thus greatly increasing the achievable level of system (product) performance, reliability, and complexity, while reducing size, cost, and power consumption.

Circuit design techniques and manufacturing technology improved, allowing the integration of more and more circuitry on a single chip. As integrated circuits became more complex, they became more specialized, losing their general-purpose nature. Many companies with large production volumes could afford to develop custom ICs designed for a specific product. The high development cost of these chips, approximately $250,000, could be justified when amortized over a production of a million or more units.

Custom integrated circuits offered many advantages. Products could be made less expensively and smaller, and they required less power. They were difficult for competitors to copy and a large portion of a product came pretested from the IC vendor. Service and assembly were simplified because fewer components were involved. Custom ICs offered marketing advantages along with performance improvements. Due to their high development cost, customized integrated circuits were not economically feasible for small companies or even large companies working on projects with a limited budget and limited production quantities until the advent of semicustom technology.

layer of metal that can make the layout of high-performance differential circuitry difficult. The more up-to-date component arrays have component groupings which are more amenable to modern design techniques. Many component arrays have been available for nearly 20 years and have several sources. They have become virtually industry standards. The amount of circuitry that can be integrated on the chip is limited by the number of components on the chip and the achievable layout efficiency for the circuitry being integrated. The amount of circuitry that can be integrated per unit of die area will typically be lower on this type of array than any other. The major disadvantages of this implementation are the relatively low level of achievable integration and the relatively high unit price.

Tile Arrays

Tile arrays are the next step in the evolution of analog semicustom technology. They consist of several miniarrays or tiles of standardized components combined into a large prefabricated array. The grouping of components on a tile is chosen to facilitate the construction of standard building block cells such as op amps, comparators, and multipliers. Although autorouting is generally not available for tile arrays, connection overlays are available to create precharacterized circuits. The components on the tile can, of course, be used like a component array to create totally customized circuitry. Tile arrays typically have two layers of metal to ease the interconnections on a tile and between tiles. The improved grouping of components and the multiple levels of interconnect give tile arrays a high degree of layout efficiency and facilitate the layout of high-performance differential circuitry. The amount of circuitry that can be integrated is limited by the number and design of tiles on the chip. The use of predesigned and characterized building block circuits significantly reduces design risks.

Standard Cell

Standard cells are much like tiles except that the cells are optimized for a particular function and they are not prefabricated. The cells exist in a CAD (computer-aided design) database and are combined as necessary to make a custom IC. The cells are connected together to yield the desired circuit function. The amount of circuitry that can be integrated on a standard cell IC is limited by economics, manufacturing limitations on die size, and the ability to interconnect the circuitry. The major disadvantage of this technology is the prototype lead time since the entire IC and all of the attendant diffusion and fabrication steps must be performed. A major advantage of standard cells over component or tile arrays is their ability to optimize performance and the level of integration. Autorouting of the metal interconnection is also available.

Fully Custom IC

A fully custom IC is designed from the ground up, just like a standard product, for the particular application at hand. All transistors, resistors, and other components and their placement on the die are designed to optimize circuit performance and minimize die size. The advantages are maximum circuit performance, minimum die size, and therefore minimum high-volume production cost. The disadvantages are a long development cycle, difficulty and expense of design changes, and a very high initial development cost.

There are migration paths through these implementations that can minimize development schedule, cost, and technical risk. For example, an ASIC may first be developed on a component array to minimize the initial development schedule and the up-front cost. Later, after the circuit and product are proven, the semicustom ASIC can be converted to a fully custom IC to minimize the high-volume unit production cost. This approach eliminates the high financial risk of initially developing the IC as a fully custom IC. Table 1-1 lists the comparative information for ASIC implementations.

Figure 1-1 illustrates the typical production volumes that justify

Table 1-1
Comparative Information for ASIC Implementations

Implementation	Lead time (weeks)	Unit price ($)	Minimum order (units)
Component array	4–8	5–12	100–1000
Tile array	8–10	7–18	1000–5000
Standard cell	10–16	7–20	5000–10,000
Fully custom	26–52	3–8	100,000

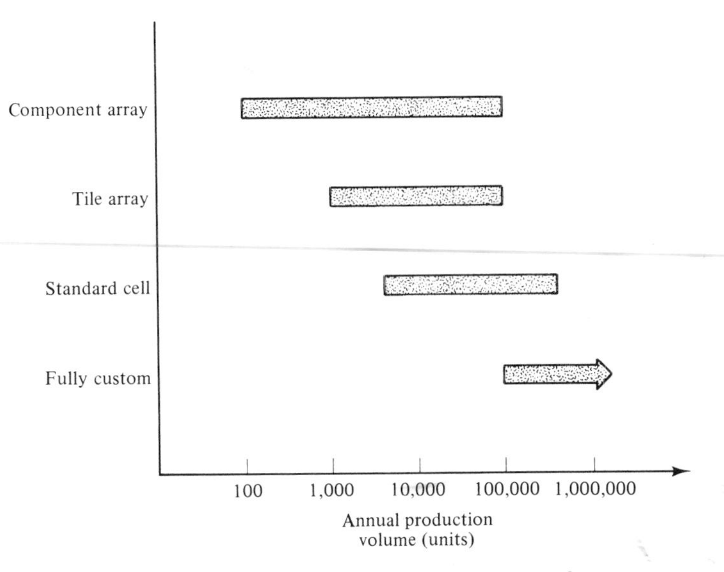

Figure 1-1 ASIC technology versus production volume.

the various ASIC implementations. The best implementation choice could be different than that indicated in the figure due to the wide range of variables involved in the choice. The figure should, however, give a first-cut estimate of the most appropriate technology.

1.2 Vendors and Product Offerings

Tables 1-2 through 1-4 list many of the product offerings from major analog ASIC vendors. It is important to realize that new products are continuously being developed and may not be reflected in the tables. This information should be a good reference for the majority of the available products and the type of products offered by each vendor. A list of major analog ASIC vendors follows.

AT&T Microelectronics Dept. 52AL330240, 555 Union Boulevard, Allentown, PA 18103, 1-800-372-2447.

Cherry Semiconductor 2000 South County Trail, East Greenwich, RI 02818, (401) 885-3600.

CLASIC (National Semiconductor) 2900 Semiconductor Drive, P.O. Box 58090, Santa Clara, CA 95052-8090, (408) 721-5000.

Exar Corporation 2222 Qume Drive, San Jose, CA 95131, (408) 434-6400.

GENNUM Corporation P.O. Box Station A, Burlington, Ontario, Canada L7R 3Y3, (416) 632-2996.

Plessey Sequoia Research Park, 1500 Green Hills Road, Scotts Valley, CA 95066, (408) 438-2900.

Micro Linear 2092 Concourse Drive, San Jose, CA 95131, (408) 433-5200.

Raytheon 350 Ellis Street, Mountain View, CA 94039-7016, (415) 968-9211.

Table 1-2
Tile Arrays

Vendor	Product	Process (V)	*npn* f_t	Total *npn*	Lateral/vertical *pnp* f_t	Total *pnp*	Special	Bonding pads
Micro Linear								
	FB308	12	720	173	12/23	79		24
	FB312	12	720	238	12/23	138		28
	FB315	12	720	307	12/23	181		44
	FB324	12	720	330	12/23	82	ECL array	28
	FB330	12	720	188	12/23	108	120 gates	42
	FB3410	36	300	132	3/20	52		32
	FB3420	36	300	286	3/20	124		46
	FB3430	36	300	394	3/20	112		66
	FB3610	12	750	178	12/24	78		24
	FB3620	12	750	268	12/24	124		32
	FB3621	12	750	329	12/24	88	ECL array	32
	FB3630	12	750	434	12/24	232		46
	FB3635	12	750	901	12/24	63	130 gates	44
AT&T								
	ALA201	10[a]	4500	68	3750	43		36
	ALA202	10[a]	4500	136	3750	86		48
	ALA300	90[a]	250	13	200	15		30
	ALA301	90[a]	250	52	200	60		32
	ALA400	33[a]	350	100	300	100		44
	ALA401	33[a]	350	68	300	68		38

[a]Consult product data.

Table 1-3
Standard Cell

Vendor	Cell type
CLASIC	Bipolar[a]
	Operational amplifiers
	Comparators
	Bias/voltage reference
	Data acquisition
	Logic
	Disk drive
	Special function
	npn geometries
	pnp geometries
	Other components
	LCMOS[b]
	Operational amplifiers
	Comparators
	Bandgap references
	Logic gates
	Special function
Total cells (bipolar and LCMOS) > 400	

[a]These standard cells are manufactured on a 15-V bipolar process with *npn* $f_t = 2.5$ GHz and *pnp* $f_t = 40$ MHz. Gate delays are approximately 1.8 ns with a fan out of three. The process also includes 2 kΩ/sq ion-implanted resistors, two layers of metal, MOS capacitors, and Schottky diodes.

[b]These standard cells are manufactured on a 3-μm double-poly process with a maximum D flip-flop toggle frequency of 50 MHz.

Table 1-4
Component Arrays

Vendor	Product	Process (V)	Total *npn*	Large *npn*	Lateral *pnp*	Vertical *pnp*	JFET	Bonding pads
Cherry								
	2000E	20	48		15			18
	2500G	20	60	2	18			18
	2800	20	60	2	25			16
	3000F	20	96	4	36			24
	3100	20	88	3	36			22
	3200L	20	88	3	36			22
	3500	20	61	2	24			22
	3600	20	123	6	52			25
	4000M	20	153	8	52	4		28
	7600	15	142	4	26			25
Exar								
	A - 100	20	60	2	18			16
	B - 100	20	69		12			16
	C - 100	20	23		8			14
	CA - 100	20	102	6	62			28
	D - 100	36	50		16			16
	E - 100	20	48		15			18
	F - 100	20	97	4	36			24
	G - 100	20	60	2	18			18
	H - 100	20	75	2	22			18
	J - 100	20	38	2	12			18
	L - 100	20	80	4	26			24
	M - 100	20	149	12	52	4		28
	X - 100	75	34	4	16			18
	U - 100	36	96		40	10	4	28
	V - 100	36	144	4	56	8	4	28
	W - 100	36	212	4	60	14	8	40
	Beta100	26	129/0		0/129			30
	Beta240	26	290/0		0/290			48

Table 1-4
(continued)

Vendor	Product	Process (V)	Total *npn*	Large *npn*	Lateral *pnp*	Vertical *pnp*	JFET	Bonding pads
GENNUM								
	LA201	20	85	2	26			24
	LA202	20	48	2	14			18
	LA204	20	27	2	9			14
	LA251	20	152	4	49			40
	LA252	20	116	4	34			32
	LA253	20	70	2	20			24
Plessey								
	MOA	20	59	2	18			16
	MOB	20	69		12			24
	MOC	20	22		8			14
	MOD	36	50		16			16
	MOE	20	48		15			18
	MOF	20	96	4	36			24
	MOG	20	60	2	18			18
	MOH	20	72	2	22			18
	MOJ	20	38	2	12			18
	MOL	20	80	4	26			24
	MOM	20	149	12	52	4		28
	MON	20	182	12	64	6		40
	MOP	20	75	4	54			24
	MOR	20	99	4	38			24
Raytheon								
	RLA40	32	37	4	17			24
	RLA80	32	46	3	19			24
	RLA120	32	39	4	16			24
	RLA160	32	43	4	10			44

CHAPTER 2

Integrated Circuit Fabrication Technology

2.0 Silicon—The Material

Man has continually used silicon and its compounds since before recorded history. Early man used them to fashion weapons and tools. In more recent times, silicon compounds have been used for everything from food containers and building material to a medium sculpted into works of art. In the last half-century, man has turned to this material of his forefathers and wrought another generation of tools to aid him in his continuing struggle to control his environment. Computers, electronic instruments, and sensors enable man to measure and predict various aspects of his environment. He can store, retrieve, combine, and manipulate vast amounts of information about things as small as an atomic particle or as vast as the universe; as simple as charge card balances or as complex as the human brain. In the electronic age, like the Stone Age, man is making good use of one of earth's most abundant resources—silicon.

Engineers, like artists, must understand the medium they are using to create their work. For this reason, a few pages are taken to review some important characteristics of silicon, the medium of integrated circuits. The basic atomic characteristics of silicon will be discussed in preparation for the section to follow on semiconductor processing. Next, techniques for modifying the characteristics of silicon will be

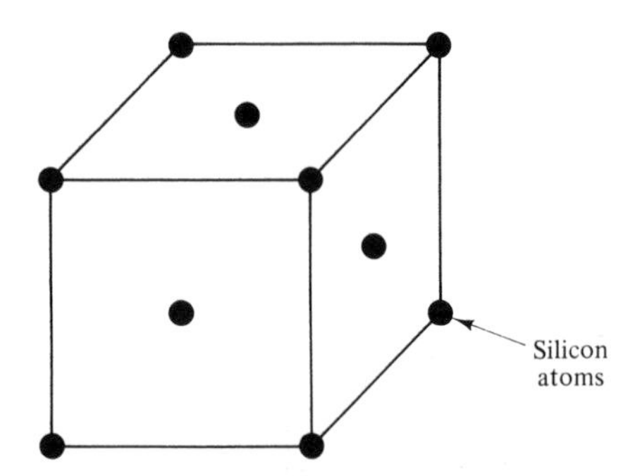

Figure 2-1 Face-centered cubic structure.

presented while pointing out strengths and weaknesses of integrated design compared to conventional circuit design philosophy.

Silicon, atomic number 14, has a face-centered cubic crystalline structure illustrated in Figure 2-1. Figure 2-2 illustrates silicon's electron configuration. There are four electrons in the outermost shell ($3s3p$), which requires eight electrons to fill. This makes silicon quadravalent. Adjacent atoms share electrons to fill the outer shell, as depicted in Figure 2-3, and are thereby covalently bonded. Silicon is

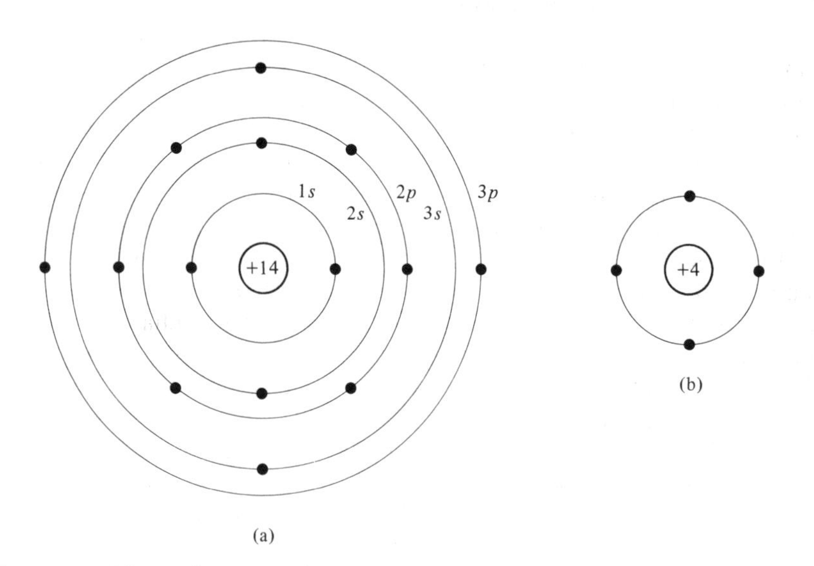

Figure 2-2 Silicon electron configuration. (*a*) Subshell population and (*b*) simplified model.

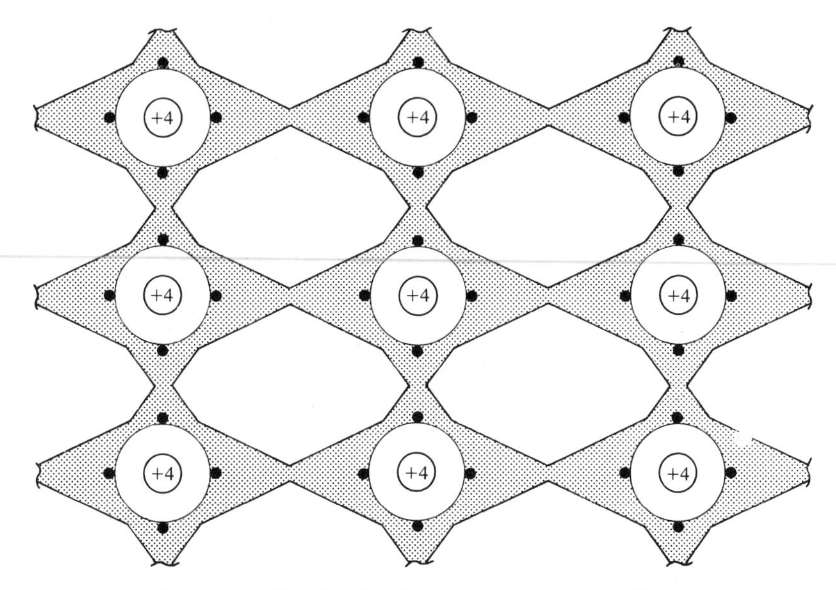

Figure 2-3 Covalent bonding in intrinsic (pure) silicon.

called a semiconductor because its ability to conduct electricity lies between that of an insulator and a conductor. A good insulator has a conductivity of 1 MΩ/cm³, a good conductor 1 μΩ/cm³, while the conductivity of silicon lies in between these values at about 50 kΩ/cm³.

Conduction in Intrinsic (Pure) Silicon

A perfect silicon crystal (i.e., one with perfectly spaced atoms such that each atom has four neighbors, each sharing four electrons) would be a perfect insulator since there would be no electrons available for conduction. Nature, however, rarely allows perfection and silicon is no exception. Structural flaws in the crystal lattice cause even the purest silicon to be made of many different crystals with discontinuities in the lattice at their boundaries. Instead of a single or monolithic crystal, silicon consists of many crystals like grains of sand. These structural defects are typically caused by contaminants but could also be caused by radiation damage.

The imperfect sharing of electrons at these discontinuities makes it relatively easy to boost the energy level of an electron from the valence band to the conduction band. When this occurs, a free or conducting electron and a positively charged electron-deficient atom are generated. The electron-deficient atom is referred to as a hole. Thus, electrons and holes are generated in pairs. The lifetime of a carrier, either a hole or an electron, is the time between its creation and the time it recombines with another electron or hole, respectively. Carrier lifetime depends upon the number of crystal defects and the number of contaminants (dopants) in the lattice. Heavy-particle radiation, such as neutron radiation, damages the crystal by causing dislocations in the lattice. These dislocations create recombination centers, which reduce carrier lifetime and increase the conductivity of the structure. Dislocations are purposely caused in some semiconductor processes by doping the crystal with gold to reduce carrier lifetime and improve switching speed.

Hole–electron pairs can be generated in a number of ways. The most common are heat, radiation (light), and the application of an electric field. Ionization energy is the energy required to cause an electron transition from the valence band to the conduction band, breaking its covalent bond. The conductivity of silicon can also be modified by mechanical stress. This last characteristic is used to make pressure transducers and can cause unexpected problems during packaging.

The conductivity of pure silicon can be modified physically with the addition of impurities and electrically through the manipulation of electric fields within the crystal. Semiconductor devices such as transistors and diodes use a combination of introduced impurities and controlled electric fields to produce gain and switching action.

Conduction in Doped Silicon

Impurities, or dopants, are added to pure intrinsic silicon to modify its conductivity. Added impurity concentrations are on the order of one part per million. Spurious impurities (contaminants) must be kept at a level of less than one part per billion or roughly a concentration three orders of magnitude lower than the desired impurities. An initial impurity concentration is added to the base or starting material during its initial crystallization process. This impurity concentration is modified by growing additional layers of oppositely doped material on the base layer, diffusing dopants into the material, or by implantation. Diffusion is accomplished by placing the silicon to be doped in a high-temperature environment rich in the desired impurity. The diffusion rate of the impurity into the silicon depends upon the concentration of the impurity, its diffusion constant, and the temperature of the environ-

once the barrier potential is canceled. The barrier potential is a negative function of temperature and therefore becomes smaller as temperature increases.

The *pn* junctions exhibit capacitance. The *p* material and the *n* material serve as the conductors, while the depletion region becomes the dielectric. It should be obvious from the previous discussion that the junction capacitance decreases as reverse bias is applied to the junction (dielectric thickness increases) and increases as forward bias is applied (dielectric thickness decreases) up to the point of conduction. Any reverse-biased junction (one that is not conducting) exhibits a capacitance that changes with the applied voltage.

2.2 Bipolar Transistor Operation

A bipolar transistor is constructed by locating two *pn* junctions contiguously in the same crystal lattice structure, as shown in Figures 2-7 and 2-8. Figure 2-7 illustrates the operation of an *npn* transistor. External bias voltages are applied to the two junctions. The emitter-base

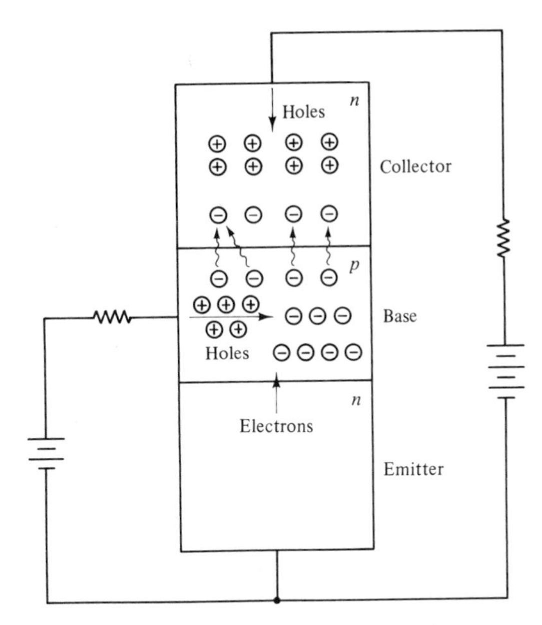

Figure 2-7 An *npn* transistor operation.

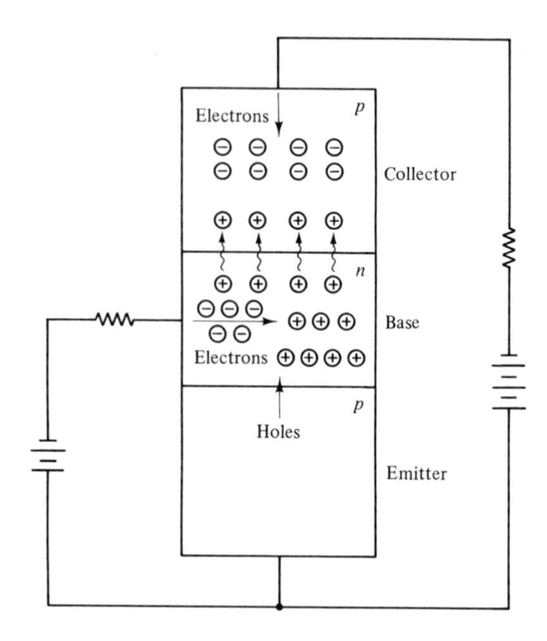

Figure 2-8 A *pnp* transistor operation.

junction is forward-biased (conducting) and the collector-base junction is reverse-biased. Majority carriers (electrons) are injected into the base region by the emitter. Some of these majority carriers from the emitter recombine with majority carriers in the base (holes). Most of the injected electrons, however, diffuse throughout the base region and are swept across the reverse-biased collector-base junction into the collector, thus forming collector current. The ratio between the base current and the collector current is referred to as beta (ß).

The operation of a *pnp* transistor is analogous to that of an *npn* and is illustrated in Figure 2-8. The operation is identical, except that the bias voltages and the roles of the holes and electrons are reversed.

2.3 Semiconductor Processing

Semiconductor processing brings manufacturing down to the atomic level. For every million silicon atoms, approximately one dopant atom is necessary to make useful semiconductor material. For every billion silicon atoms, less than one spurious contaminant atom can be allowed without adversely affecting the characteristics of the material. Thus, the manufacturing process must be able to regulate the numbers and purity of atoms to a remarkable degree.

The raw material for this technology is lightly doped *p*-type silicon. Different crystallographic orientations are used (i.e., Miller index 111, 100, etc.), depending on the process. An ingot of monolithic silicon is grown from a seed crystal (see Figure 2-9) by slowly pulling it from a vessel of molten silicon. As the silicon cools, it crystallizes into a monolithic structure. The pull rate largely determines the diameter of the ingot, which ranges between 3 and 8 in. The ingot is next sliced into wafers approximately 500-μm thick. The wafers are polished, inspected, and packaged for shipment to the IC manufacturer, ready for processing. The flat on the wafer is used for orientation of the wafer during processing.

The basic manufacturing process consists of several steps that are repeated until the desired structures are completed. The steps are oxidation, photomasking, etching, doping, and diffusion. After the final diffusion is complete, one or more oxidation, photomask, etch, metal, etch passes will be made, depending on how many layers of metal interconnection (usually 1 to 2) there are. These final steps are used to interconnect the components on the chip and create the circuit.

Semicustom arrays, either the random component type or tile ar-

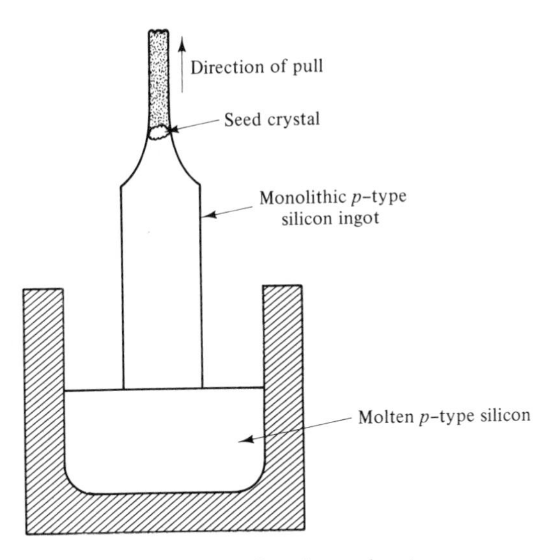

Figure 2-9 Growing an ingot.

rays, are processed through the final diffusion and contact etch, coated with the first layer of metal, and then stored. At this stage of manufacturing all of the components have been created on the chip and shorted together but have not been interconnected to form a circuit. The die is analogous to a printed circuit board stuffed full of various useful components but without the copper lands defined.

Before any processing can take place, the components must be designed. Transistor geometries, resistor lengths and widths, and so on, must be defined. This information is then used to determine the size, shape, and location of each diffusion required to construct these devices. The dopant concentration for each layer is predefined for each "process." Each process has a list of design rules that specifies the minimum feature sizes and the spacing between each layer of the process. Special test die are placed on the wafer to monitor the process tolerances. Data gathered from the test die is used to fine-tune the manufacturing process.

The design rules are determined by the characteristics of the process. A typical differentiator between processes is maximum breakdown voltage. The higher the maximum breakdown voltage, the wider the spacing required between diffusions. Higher reverse bias means that junction depletion spreads will be wider. Also, lighter dopant concentrations must be used to prevent the development of a critical field and breakdown prior to the specified limit. In summary, the higher the maximum breakdown voltage for a process and the wider the spacings between components and diffusions, the less component density the chips will have.

The mask for each layer of an integrated circuit is defined using a computer-aided design (CAD) system. The graphic representation for each opening in every layer of the process is entered into the computer. A program called a design rule checker (DRC) is then run against the database from the layout created in the CAD system to ensure that no design rule violations are present in the mask set. The masks are now ready to be used to define the diffusions in the starting material.

The integrated circuit manufacturing process has very little to do with electronics. It is a combination of high-resolution photolithography, high-precision electron beam technology, metallurgy, chemistry, and material science. The transistor, resistor, and diode structures are literally grown with the crystalline structure of the finished wafer. First we will take a detailed look at the oxidation, photomasking, etching, doping, and diffusion operations used to manufacture integrated circuits. Next we will step back and look at the typical layers that are used to create an integrated circuit and follow this process, omitting most of

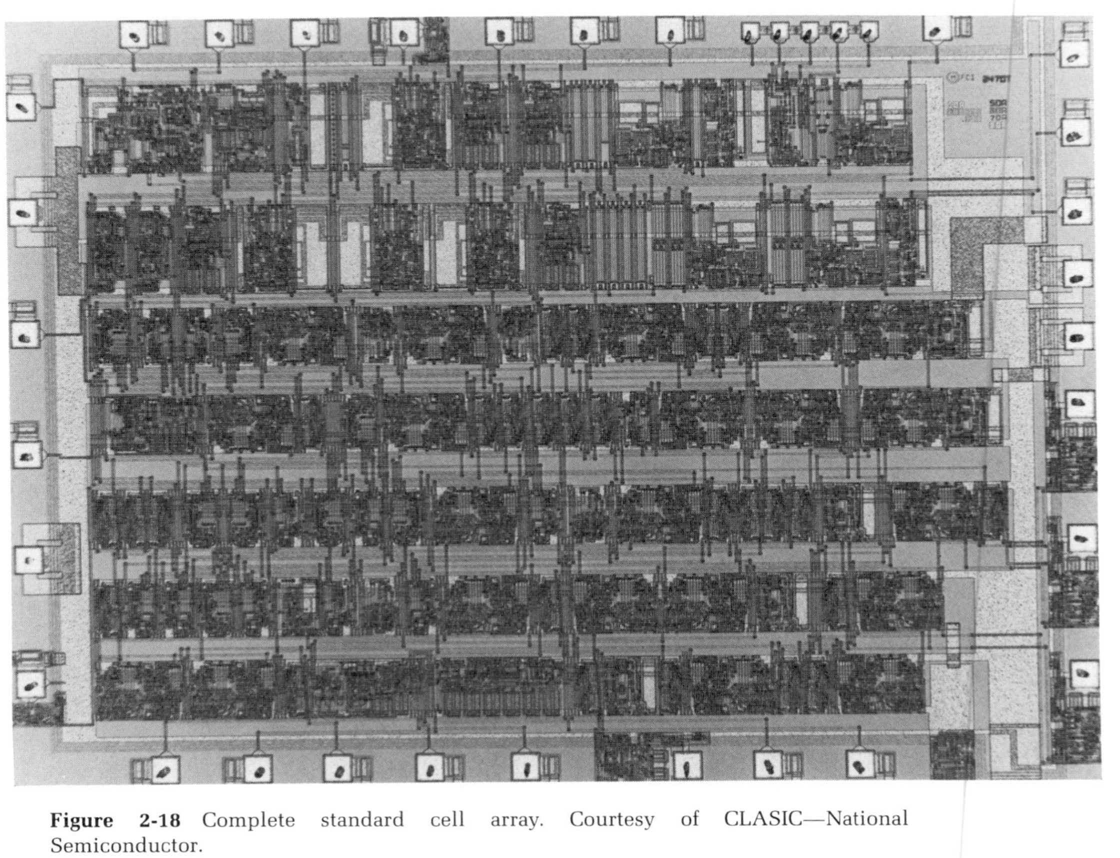

Figure 2-18 Complete standard cell array. Courtesy of CLASIC—National Semiconductor.

within specification. If the wafers pass, they are sent to the first operational test, called die sort.

Automatic probes step across the wafer one die at a time, testing the functionality of each die. This test is meant to reject any die that are nonfunctional or grossly out of specification. Rejecting bad die at this point saves the unnecessary expense of packaging. Bad die are marked with an ink dot so they can be easily identified and discarded when the die are separated from the wafer.

After die sort, the wafer is reduced in thickness from 500 to approximately 200 μm with a procedure called backlap, and the die are separated with a diamond saw. The nonfunctional or marginally functional die are identified with an ink dot during die sort and discarded. The good die are visually inspected for scratches, chips, or other physical defects. The remaining die are now ready to be packaged.

The packaging process consists of five basic steps: die attach, lead bonding, seal (encapsulating for plastic packages), lead plating, and lead forming. Several inspections are made during the packaging process to ensure that it is being done correctly.

After the ICs are packaged, they are ready for final test. This is the last electrical test the devices will receive prior to shipment. This testing can be relatively simple or very complex, depending on the requirements of the customer. The tests can be conducted over a range of temperatures and can include complex high-frequency measurements. The tailoring of these tests is part of the "customization" of the ICs. Obviously, the more detailed the testing, the more expensive it will be. Good circuit design can ensure specification compliance while minimizing complex test requirements. More on this subject in a later chapter.

The "good" parts are finally marked with a part number and packed for shipment to the customer.

References

1. Grove, A.S., *Physics and Technology of Semiconductor Devices.* Wiley & Sons, New York, 1967; p. 37.
2. Grove, Chapters 1–3.

for abrupt junctions such as very shallow diffused junctions or those formed by ion implantation, and

$$C_j = \frac{C_{j0}}{\sqrt[3]{1 - \frac{V_b}{\Psi_0}}}$$

for a linearly graded junction such as those formed by most standard diffusion processes,[5] where

C_j is the junction capacitance dependant on V_b
C_{j0} is the junction capacitance with $V_b = 0$

The *npn* collector-base junction will appear to be a graded junction for $V_b < 1$ V and an abrupt junction for larger reverse biases. The *npn* emitter-base junction typically appears to be a linearly graded junction. These equations are valid in the forward-biased condition (i.e., $-V_b = \Psi_0$) until forward current begins to flow.

There are two significant points: (1) junction capacitance increases rapidly as $-V_b$ approaches Ψ_0. Since Ψ_0 is directly proportional to temperature, C_j is sensitive to temperature. (2) Significant changes in C_j can occur due to temperature changes if $-V_b \approx \Psi_0$ (be careful if C_j is being used as frequency compensation to stabilize an amplifier).

Forward-Biased *pn* Junctions

The forward current in a *pn* junction is described by the following equation:

$$I = I_s[\exp\left(\frac{q\phi}{kT}\right) - 1]$$

Several points about this equation are noteworthy:

1. The -1 is insignificant and can be omitted in most practical cases.
2. k and q are constants.

3. I is directly proportional to I_s.
4. I is proportional to the inverse exponential of T.

If this equation is solved for ϕ in terms of I, the forward voltage drop of a *pn* junction is described as

$$\phi = \frac{kT}{q} \ln \frac{I}{I_s}$$

where

k = Boltzmann's constant (8.62 E-5 eV/K)
T = absolute temperature (T_a + 273°C)
q = electron charge (1.60 $E-19$ coulomb)
I_s = junction reverse saturation current
I = the forward current flowing through the junction when the voltage drop is measured

Several points about this equation are noteworthy:

1. k and q are constants.
2. I_s is a function of the dopants, their relative concentrations, and geometric characteristics of the junction.
3. The forward voltage drop ϕ is directly proportional to absolute temperature.

These characteristics can be used to perform an amazing array of precision analog functions. Examples will be presented in Chapter 5.

Junction Breakdown

A reverse-biased *pn* junction will exhibit a small current flow due to minority carrier holes and electrons that are accelerated across the depletion region by the electric field (there could also be leakage contributed by ionic contamination on the surface of the wafer). This leakage current will typically be low enough in silicon that it can be ignored for all but the most demanding circuit designs. This leakage current, however, doubles for every 10°C rise in junction temperature.

As the reverse bias is increased, the carriers are accelerated with more and more energy until a critical electric field is established across the depletion region. At this point, the carriers are accelerated to a high enough energy to create additional electron-hole pairs through collisions with silicon atoms. After this point, current increases dramatically with a small increase in voltage and current must be externally limited to avoid excessive power dissipation and the destruction of the junction. This chain-reaction process is called "avalanche breakdown." The voltage at which this critical field occurs depends on the doping concentrations of the two materials. The higher the concentrations, the lower the breakdown. The depletion spread in relatively heavily doped regions is small, thus allowing a much stronger electric field to be established across the shorter depletion region with a lower applied voltage.

Emitter-base junctions of *npn* transistors are commonly used as voltage references since they break down at approximately 6.5 V. This diode, when used in this manner, is referred to as a Zener diode. This is a misnomer since the breakdown mechanism is avalanche. The Zener breakdown process occurs at low voltages in richly doped material and is not typically found on semicustom ICs.

3.2 Resistors

There are three types of resistors commonly found in analog semicustom arrays: base-diffused, ion-implanted, and pinched. Each resistor type has characteristics that make it best suited for certain applications. It is important to remember that the resistors found on integrated circuits are junction-isolated devices. They consist of a *p*-type material diffused or implanted into an *n*-type epi island. To conserve die area, many resistors are typically placed in a common epi island. It is important, therefore, to ensure that the *n*-type epi island is at a potential equal to or higher than any potential that will appear on any resistor. If this condition is violated, many parasitic effects can occur. These undesirable effects are described in more detail in Section 3.8.

Base-Diffused Resistors

Base-diffused (base) resistors are the most commonly used resistors in integrated design. Their values on semicustom arrays typically range from 200 Ω to 5 kΩ. In custom applications, or standard products, base resistors have been routinely made in the range of 50 kΩ. There is, however, a significant die area penalty for these large values. The absolute values of base resistors can vary by as much as ±30% from wafer to wafer due to variations in the base diffusion sheet resistivity and masking and etching tolerances. The resistors match and ratio to within 1 to 2% on a given die. The breakdown voltage for these resistors is the same as the *npn* collector-base breakdown for the process they are fabricated on. Base resistors have relatively good high-frequency performance but are limited somewhat by the base-epi parasitic capacitance. They have a relatively low voltage coefficient compared to ion-implanted resistors.

The geometry for a typical base resistor is shown in Figure 3-2. These resistors are also referred to as "dog bone" resistors because of their shape. The *p*-type region around the contact does not have to be larger than the resistor body and can have a shape such as that in Figure 3-3. Wide resistors are less sensitive to masking and etching toler-

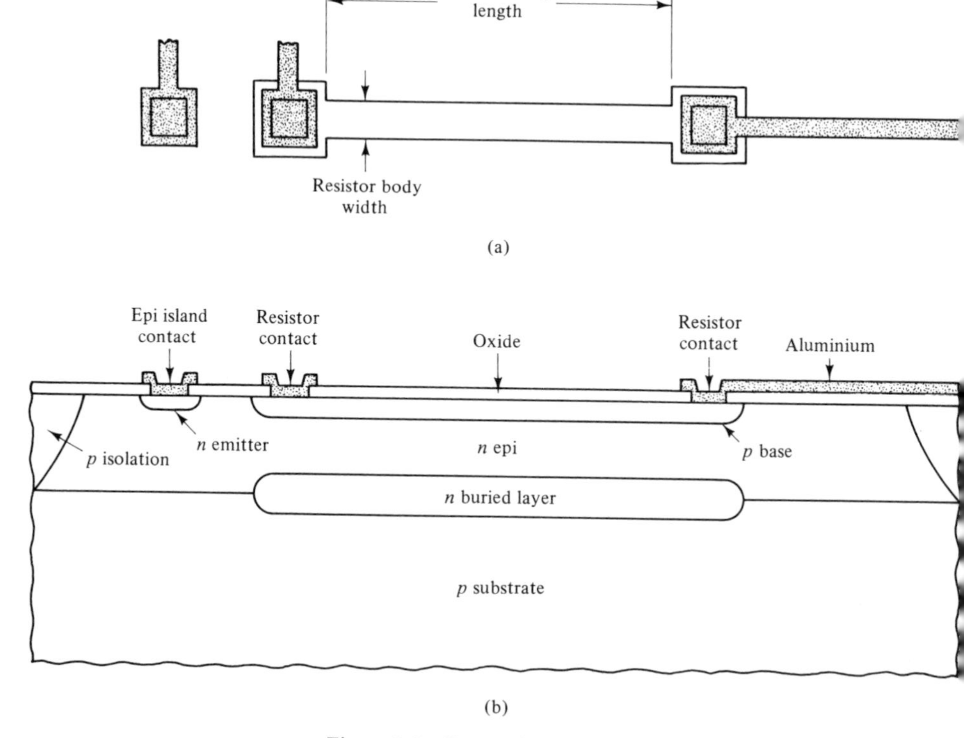

Figure 3-2 Base resistor geometry.

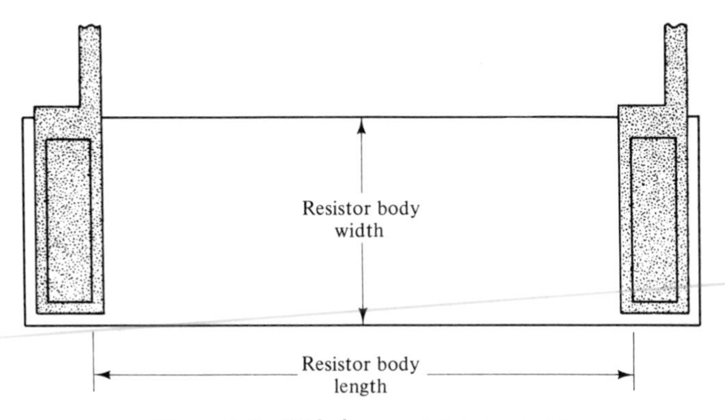

Figure 3-3 Wide-base resistor geometry.

ances and also exhibit better matching. For a given value, wider resistors must be longer and therefore take up more die area.

The value of a base resistor is determined by the length-to-width ratio with a minor contribution from contact resistance. Contact resistance is important in small value resistors only.

$$R = \frac{L}{W}\rho_{sb} + 2R_c$$

where

R = resistance in ohms
L = resistor body length
W = resistor body width
ρ_{sb} = base diffusion sheet resistivity in Ω/sq
R_c = contact resistance

Base sheet resistivity typically ranges between 100 and 200 Ω/sq and contact resistance from 2 to 10 Ω. Minimum resistor widths depend on the vendor and process.

A serpentine structure (Figure 3-4) is used to make very long and therefore high-value base resistors. This structure allows a very long resistor to be compressed into a relatively compact area. This layout technique improves the matching of large-value resistors (assuming they are either intertwined or closely spaced) and minimizes any differences in resistor values due to thermal gradients across the die.

The addition of corners to the body of the resistor increases its

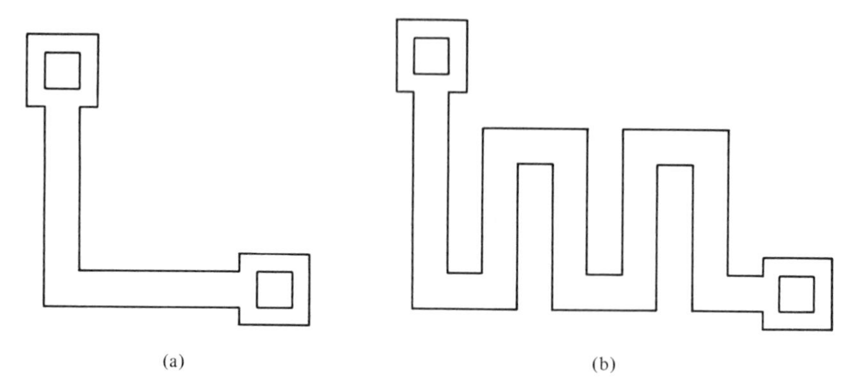

Figure 3-4 Resistors with corners.

resistance due to current crowding at the corners. This means that the effective width of the resistor at the corner is less. The value of this type of resistor is

$$R = \left(\frac{L}{W} + N_c M_c\right)\rho_{sb} + 2R_c$$

where

R = resistance in ohms
L = resistor body length
W = resistor body width
ρ_{sb} = base diffusion sheet resistivity in Ω/sq
R_c = contact resistance
N_c = number of corners
M_c = corner coefficient

The corner coefficient ranges from 0.25 to 0.55 with a linear slope depending on resistor width and junction depth. Narrow resistors have a lower coefficient value. A typical value for M_c would be approximately 0.30.

The value of a base resistor and its associated capacitance to the *n* epi in which it is embedded depends on the bias of the resistor with respect to the epi. It is critical to ensure that this is a reverse bias to maintain the independence of resistors. The amount of reverse bias will then modulate the depletion spread and thus the cross-sectional area (mostly the depth of the resistor), changing the resistor's ohmic value and the value of junction capacitance. This effect is illustrated in Figure 3-5. Additionally, as current flows through the resistor and voltage is dropped across it, the reverse bias will increase along the length of the resistor toward the low potential end. The capacitance of the resistor to the epi will also modulate inversely with the reverse bias. Both the resistor value and associated capacitance will vary with applied bias and signal swing. For base resistors, the voltage coefficient is on the order of $\pm 0.01\%/\mathrm{V}$. This value is much higher for ion-implanted resistors due to their shallow junction depth. The variation of capacitance depends on the value (and size, i.e., junction area) of the resistor and is usually important only for high-gain, high-frequency amplifiers. Table 3-1 summarizes typical diffused resistor characteristics.

Figure 3-6 illustrates a base-diffused resistor and the major parasitic elements associated with it in the normal operating mode. The resistor-epi diodes and their associated junction capacitance are actually distributed along the length of the resistor. These components are shown in the figure as lumped parameters for simplicity.

Under normal bias conditions the *p*-type substrate is at the lowest potential on the chip, and the *n*-type epi island is connected to a potential greater than the highest potential that will appear on any resistor in that epi island. In most cases the epi island is connected to the positive supply voltage unless there is some constraint, such as metal routing, that prevents it.

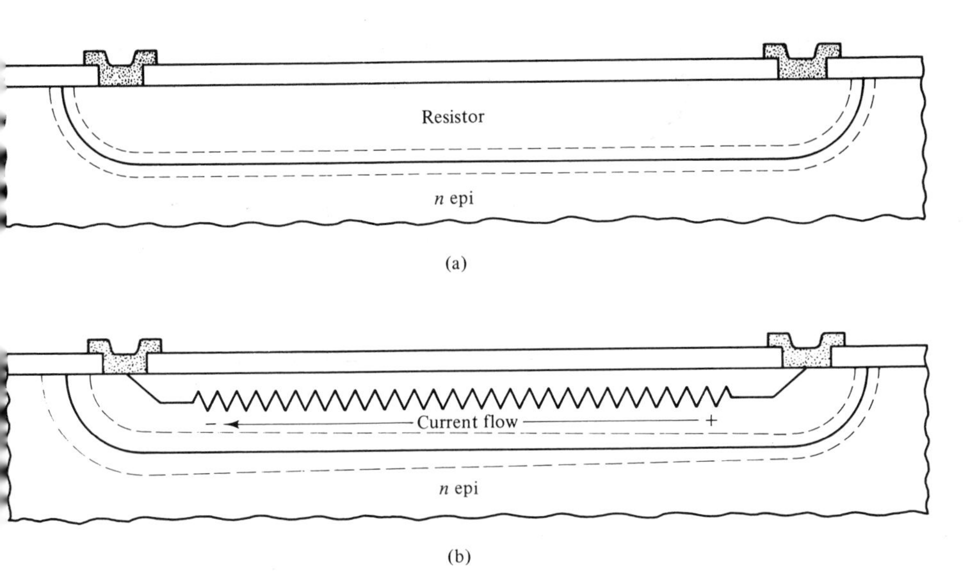

Figure 3-5 Depletion spread error on resistor.

Table 3-1
Base Resistor Typical Characteristics

Parameter	Typical values
Value range	100 Ω–5 kΩ
Absolute value tolerance	±20%
Temperature coefficient	
−55°C to −25°C	−650 ppm/°C
−25°C to 0°C	+150 ppm/°C
0°C to 25°C	+700 ppm/°C
25°C to 75°C	+1000 ppm/°C
75°C to 125°C	+1500 ppm/°C
Voltage coefficient	±0.01%/V
Matching (identical geometries in close proximity)	±2%

Ion-Implanted Resistors

Ion-implanted resistors are similar to base-diffused resistors except that they have a very shallow junction depth. This gives them a very high (1 to 5 kΩ/sq) sheet resistivity, thus allowing high value resistors to be fabricated in small areas. They do, however, have a significant voltage coefficient (about ±0.2%/V). Ion-implanted resistors are, in effect, JFETs without a top gate and have a nonlinear voltage coefficient. This effect is much worse for narrow resistor geometries. The equations

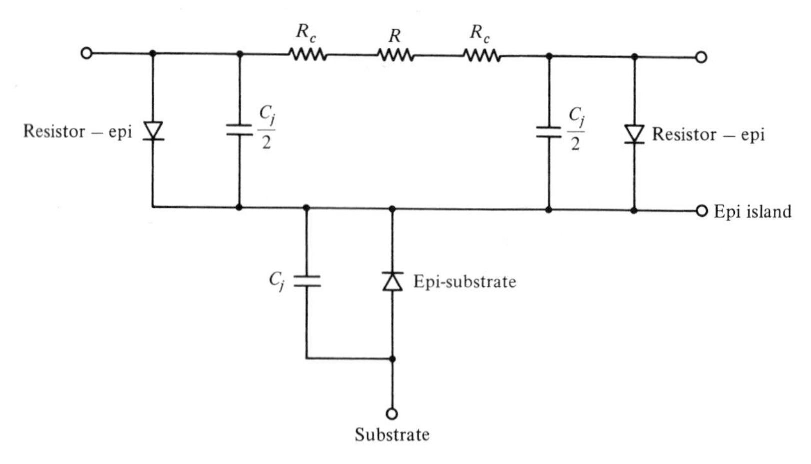

Figure 3-6 Base resistor equivalent circuit.

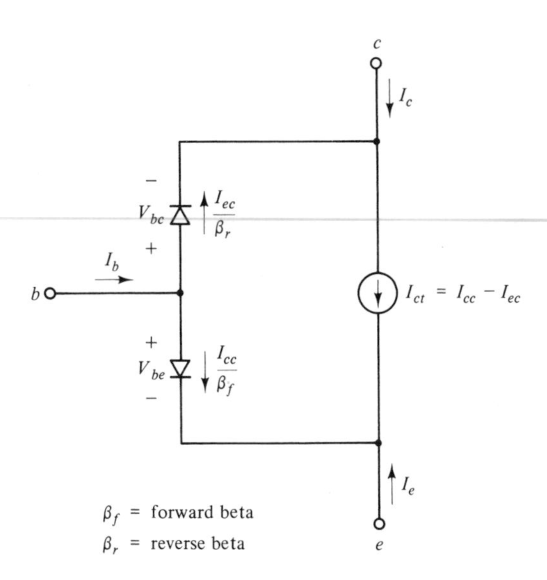

Figure 3-9 Simple dc Ebers–Moll model.

This model contains no provision for the bulk or contact resistance associated with the transistor terminals. Charge storage is not modeled and both forward and reverse beta are considered constants. It is suitable for dc analysis or for a quick bias analysis of complex circuitry. The major advantage of this model is computational efficiency.

A more accurate model with terminal resistances and capacitances added is illustrated in Figure 3-10. This is not only a more accurate dc model but provides for the simulation of charge storage effects. The unprimed terminals are the "touchable" terminals and the primed terminals are the active terminals. The added components are as follows:

$r_{bb'}$ = Resistance from base contact to active base region
$r_{cc'}$ = Resistance from collector contact to active collector region
$r_{ee'}$ = Resistance from emitter contact to active emitter region
C_{Dc} = Collector-base diffusion capacitance
C_{jc} = Collector-base junction (depletion) capacitance
C_{De} = Emitter-base diffusion capacitance
C_{je} = Emitter-base junction (depletion) capacitance
C_{cs} = Collector-substrate (epi-substrate) junction (depletion) capacitance

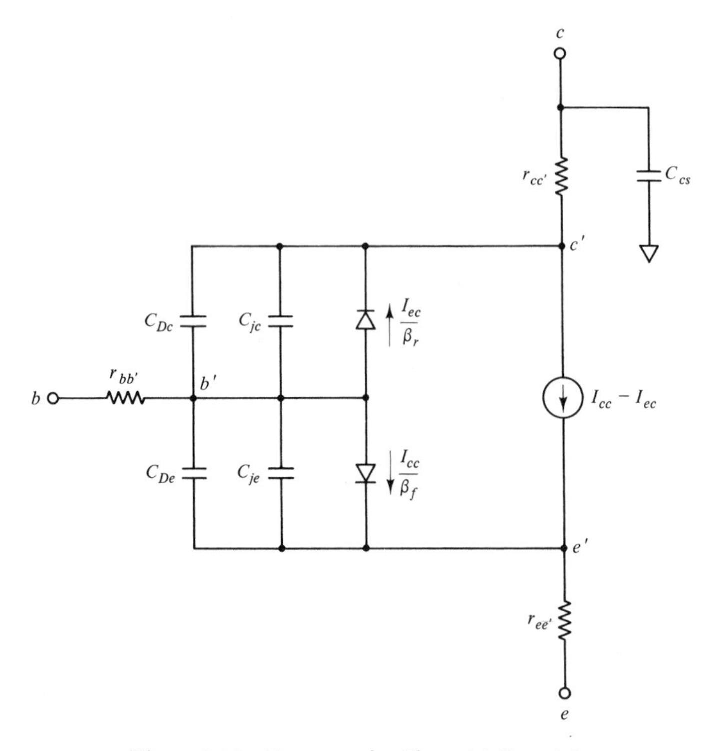

Figure 3-10 More complex Ebers–Moll model.

Hybrid-π Model

The small-signal "linearized" version of the model, illustrated in Figure 3-11, is the well-known hybrid-π model. In this model, the components are determined as follows:

$$
\begin{aligned}
r_{bb'} &= \text{(same as for Ebers–Moll)} \\
r_{cc'} &= \text{(same as for Ebers–Moll)} \\
r_{ee'} &= \text{(same as for Ebers–Moll)} \\
r_{b'e} &= kT/qI_e \\
r_{b'c'} &= kT/qI_{ec} \\
C_{b'e'} &= C_{De} + C_{je} \\
&= C_{Dc} + C_{jc} \\
C_{cs} &= \text{(same as for Ebers–Moll)}
\end{aligned}
$$

The component $r_{b'c}$ is used to simulate collector-base leakage I_{ec} and is frequently assumed to be infinite. A detailed derivation of these models can be found in Reference 6.

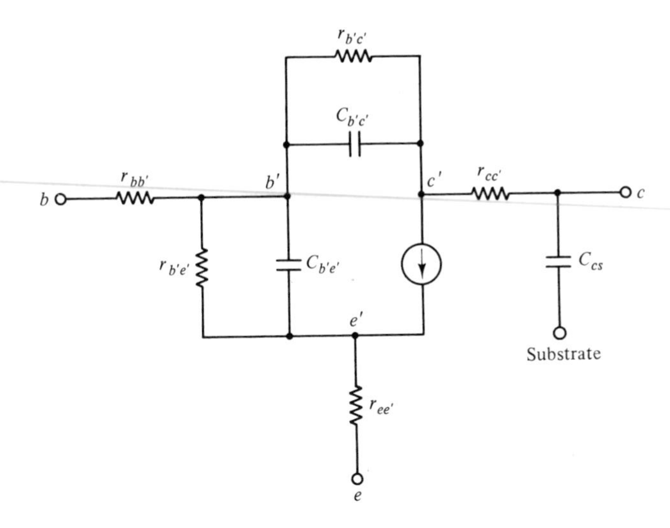

Figure 3-11 Hybrid-π model.

Small-Signal *npn*

Typical small-signal *npn* geometries are shown in Figure 3-12. Small-signal *npn* geometries can have either one or multiple collector contacts. The multiple-collector contact geometries allow for greater layout flexibility. A cross-sectional view of a small-signal *npn* with the model elements included is shown in Figure 3-13. Inspection of the figure gives a good intuitive feel for some of the limitations of this device. The component $r_{ee'}$ is the combination of the contact resistance between the

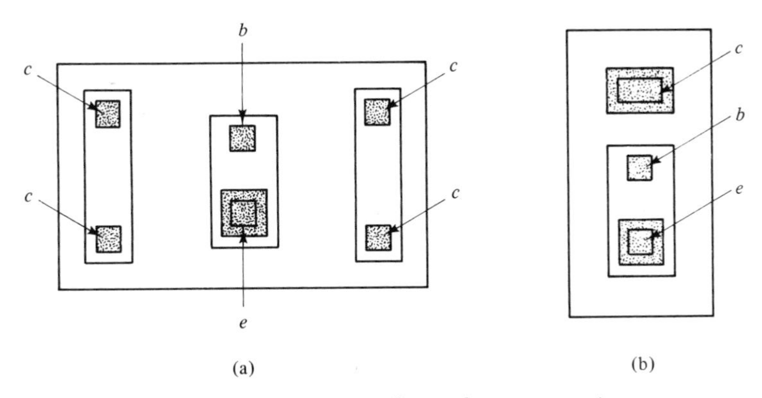

Figure 3-12 Typical small-signal *npn* geometries.

$$C_{cs} = C_{\text{epi-iso}} + C_{\text{epi-substrate}} + C_{\text{buried layer-substrate}}$$
$$R_{c'c} = R_c + R_e + R_{\text{epi}\,1} + R_{\text{bl}} + R_{\text{epi}\,2}$$
$$R_{bb'} = R_c + R_b + R_p$$
$$R_{e'e} = R_c + R_e$$

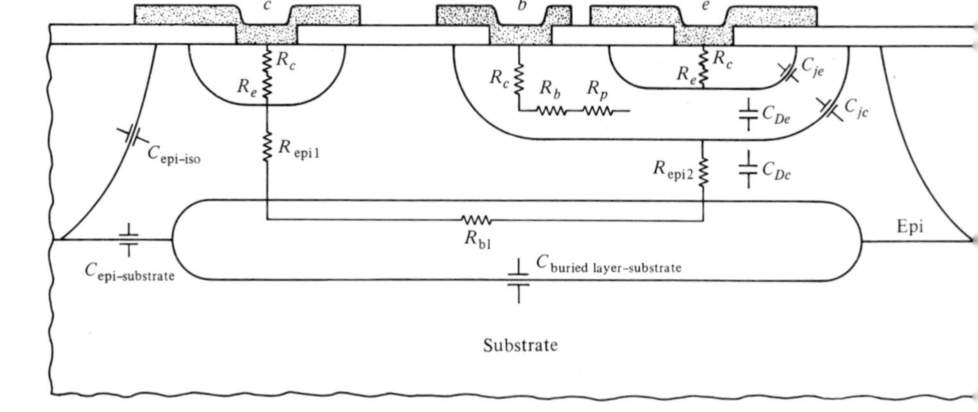

Figure 3-13 Small-signal *npn* cross section with model elements.

aluminum and the *n*-emitter material and the bulk resistance of the *n* emitter between the contact and the active emitter area. Although these values are typically small they can contribute to the overall saturation resistance of the device when used as a switch. When $r_{ee'}$ is multiplied by $\beta + 1$ to be reflected into the base circuit, its value becomes 100 to 300 times greater and can impact the high-frequency response of an amplifier circuit.

The collector resistance $r_{c'c}$ is the sum of the contact resistance R_c; the bulk resistance of the *n*-emitter diffusion used to contact the epi, R_e; the bulk resistance from the *n*-emitter diffusion to the buried layer, R_{epi1}; the resistance of the buried layer, R_{bl}; and the resistance from the buried layer to the active base area, R_{epi2}. It is readily apparent from this analysis why the heavily doped, low-resistivity, buried-layer diffusion was added to the *npn* geometry. Without it, the collector saturation resistance would be extremely high. The collector saturation resistance is much lower on discrete transistors because there is no substrate and the collector contact is made directly under the base region.

The base resistance is the combination of the contact resistance R_c, the bulk resistance R_b of the base diffusion, and the resistance of the base region which is "pinched" by the emitter diffusion. The pinched component R_p is created by the same mechanism as that used to create a pinched resistor. This component of resistance is modulated

ated. The emitter and collector of the lateral *pnp* are formed from the *npn* base, which is not a very rich diffusion. This limits the *pnp* emitter efficiency. The required spacing between two base diffusions (emitter to collector in a lateral *pnp*) is very large (approximately 0.6 mils), which makes the effective base width very wide, increasing the chance of carrier recombination in the base region. This increases the base current for a given collector current, resulting in a relatively low beta.

The emitter-to-collector current flow is lateral, as the name implies. The purpose of the buried layer in a lateral *pnp* is to keep carriers from being collected by the reverse-biased, epi-substrate junction. Some collection, however, does occur at the epi-isolation junction. For this reason substrate contacts should be made in relatively close proximity to lateral *pnps* to prevent substrate currents from flowing long distances. Since the transistor action is lateral, the effective emitter area is not the area of the diffusion (as in the case of an *npn*) but rather the area around the periphery of the emitter.

Figure 3-18 shows the model elements in a lateral *pnp*. The Early voltage for lateral *pnp* transistors is typically 80 to 100 V. Their output impedance is therefore one-half to one-third that of *npn* transistors. This fact has a significant impact on realizable gain when using lateral *pnp* transistors as current source collector loads or as gain stages. Table 3-6 gives typical lateral *pnp* characteristics.

Vertical *pnp* Transistors

Vertical *pnp* transistors are similar to laterals in that a base diffusion is used for the emitter and the epi is the base. The difference is the omission of the buried layer; the epi-substrate and epi-isolation junctions serve as the collector-base junction. The emitter-to-collector current flow is mostly vertical. Although the emitter efficiency is the same as that for a lateral *pnp*, the base width is significantly less. Vertical *pnps*, therefore, have higher beta at higher currents than laterals and higher f_t. The major peculiarity of vertical *pnps* is that their collectors are committed to the substrate, which must be connected to the negative power supply. Vertical *pnps* are typically used in the lower half of push–pull output stages or as emitter followers for level shifting applications. A major concern about using vertical *pnps* is to make very sure that substrate contacts are made in close proximity. This is especially true for devices that are used in output stages. The higher currents associated with vertical *pnps* can cause unexpected trouble if allowed to flow long distances through the substrate.

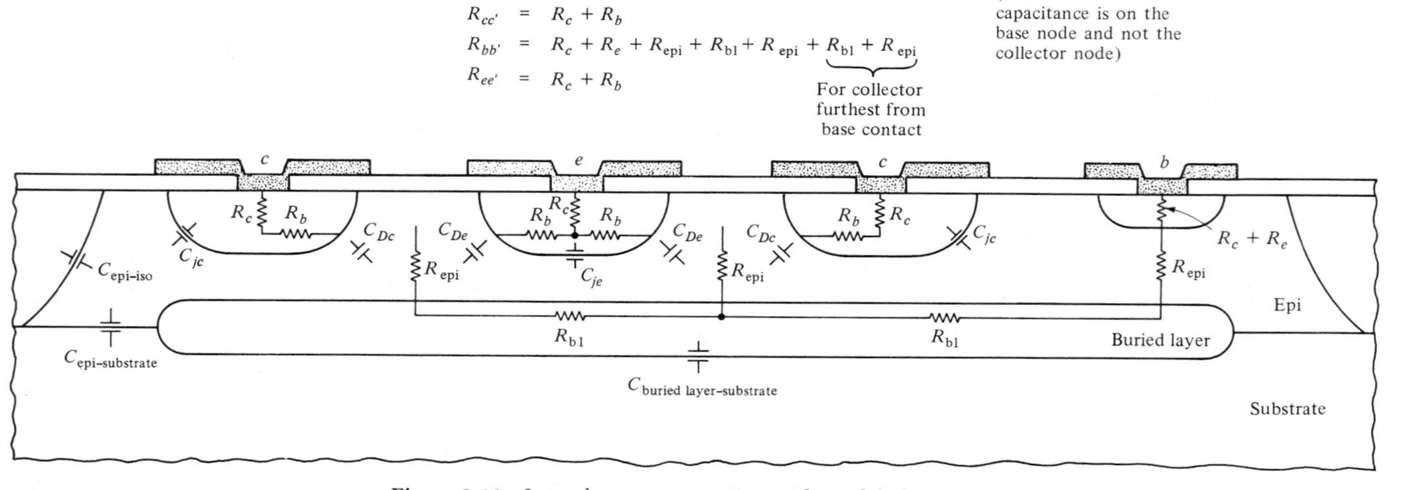

Figure 3-18 Lateral *pnp* cross section with model elements.

Table 3-6
Typical Lateral *pnp* Characteristics

Parameter	Typical value
β	50
β matching	5%
f_t	4 MHz
R_{sat} (saturation resistance @ 10 mA)	500 Ω
I_{ceo}	2 nA
V_{be} matching	2 mV
BV_{ebo}	Maximum process voltage

3.4 Diodes

Diodes do not exist as separate components on semicustom ICs but are readily made by connecting the base and collector of a transistor. This is not as area-efficient as having a dedicated diode geometry. However, the small amount of wasted silicon is more than made up for by the layout flexibility gained by having transistors play a dual role. The *npn* and lateral *pnp* transistors are most frequently used to make diodes. Vertical *pnp* transistors could be used for some specialized clamping function but are very limited since their collector (cathode) is committed to the negative supply (substrate). Using the terminals of any other component, such as a resistor, to make a diode is highly undesirable. This will almost certainly activate parasitic components that will interfere with the normal operation of the IC.

The *npn* and *pnp* transistors used as diodes have their own peculiar characteristics. The *pnp* diodes have a high reverse breakdown voltage equal to the *npn* transistor's collector-base breakdown voltage. The *pnp* diodes cannot handle large currents and tend to lose current to the substrate due to the parasitic collector action of the epi-isolation junction. The *npn* diodes, on the other hand, have a reverse breakdown equal to BV_{ebo}, which is approximately 6.5 V. This effect can be a problem when using the diode as a rectifier but is very useful when a Zener diode is needed. Be sure that the collector and base of an *npn* are connected when using the device as a Zener. Never allow the base to float as is sometimes done in discrete designs. This can give rise to undesirable parasitic effects. Diodes can handle the same amount of current as the transistor from which they are made. Most semicustom vendors provide separate curves and data for transistors used as diodes. More details will be given on the use of diodes in Chapter 5 in the course of describing integrated circuit design concepts.

3.5 Junction Field-Effect Transistors

Ion-implanted *p*-channel junction field-effect transistors, JFETs, are the newest active devices to be added to bipolar analog semicustom arrays. The JFETs available on semicustom arrays, however, are usually fairly small-geometry devices (relatively high R_{on}, and low I_{DSS}). They are very similar to ion-implanted resistors except an *n*-type implantation is made over the *p* implantation to serve as the top gate of the JFET. Figure 3-19 illustrates a typical JFET geometry found on semicustom arrays. These devices are very useful for many applications including high-input impedance buffer amplifiers, voltage-controlled attenuation, analog switching, current sources, and gain switching, to name just a

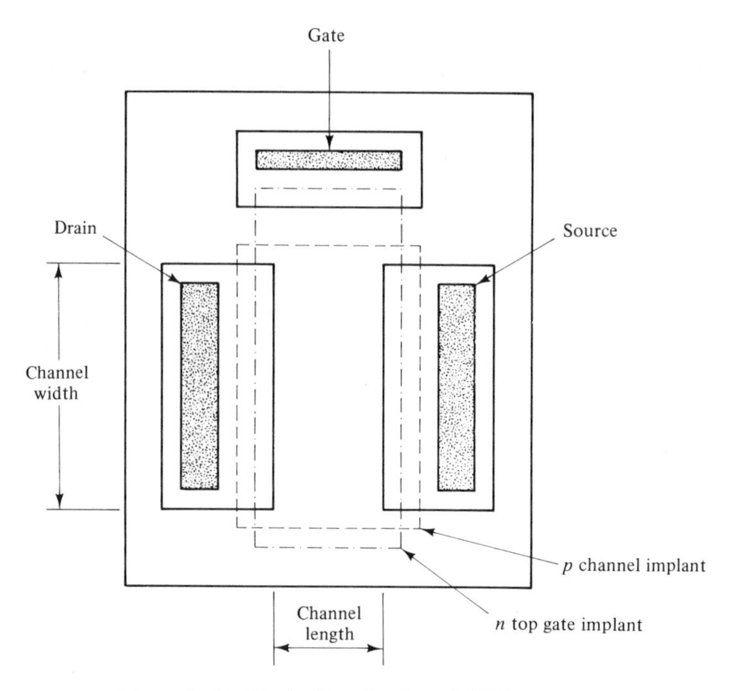

Figure 3-19 Typical ion-implanted JFET geometry.

few. While the JFET will not supplant the bipolar transistor on semicustom arrays, it is a useful addition to the analog designer's toolbox.

First the model of a JFET will be reviewed, then the actual physical structure of the device will be discussed. The model elements will then be related to the structure of the device. This will help to cement the concepts of the model with the physical structure of the device and lay the groundwork for SPICE modeling in Chapter 4 and circuit design concepts in Chapter 5.

JFET Model

The JFET is a depletion-mode device (it is on until turned off) and behaves like a voltage-controlled resistor. The ends of the "resistor" (channel) are referred to as the source and drain and are interchangeable. The control element is referred to as the gate. Figure 3-20 illustrates the action of a JFET. As the gate-to-source voltage, V_{GS}, becomes increasingly positive the conducting channel thickness is depleted until $V_{GS} = V_p$ (V_p is the pinch-off voltage) is reached and the channel is completely depleted (pinched off). The gate modulates the resistance of the channel by changing its thickness and therefore its resistivity.

Current flowing through the device also modifies the resistance of the channel. As current flows from the source to the drain, a voltage drop is produced along the channel, causing it to deplete. Eventually this action becomes self-limiting and the JFET becomes a current source. This limiting current is referred to as I_{DSS} if the gate and source

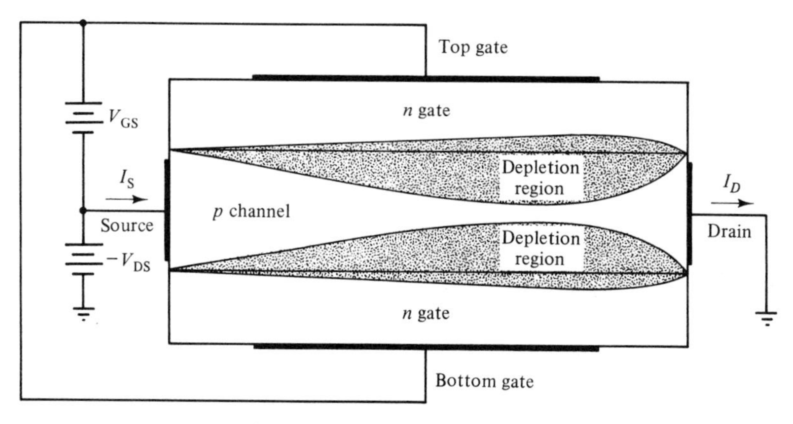

Figure 3-20 JFET operation.

are connected together (i.e., $V_{GS} = 0$). Given enough voltage between the source and drain, the JFET can serve as a voltage controlled current source with a range of values from I_{DSS} to a very much lower value limited by junction leakage.

Figure 3-21 is the schematic of a p-channel JFET model. C_{GS} is the gate-source depletion capacitance associated with the gate-source diode, D_1, and C_{GD} is the gate-drain depletion capacitance associated with the gate-drain diode, D_2. R_S and R_D represent the ohmic source and drain resistance, respectively. When the FET is used as a low current ($I_D \ll I_{DSS}$) switch with $V_{GS} = 0$, $R_{on} = R_S + R_D$. As the current increases, R_{on} also increases due to the voltage drop in the channel and the resulting increase in depletion spread into the channel. As the current continues to increase, the resistance increases until the current is limited at I_{DSS}.

Figure 3-22 illustrates a set of characteristic curves for a p-channel JFET. Notice that the I_D versus V_{DS} curve is linear for low values of I_D and V_{DS}. This is the region in which JFETs are used for analog switches. The on resistance is the inverse slope of the line in the linear portion of the curve. Also notice that the slope of the linear portion of the curve changes with increasing V_{GS}. The on resistance is the lowest at $V_{GS} = 0$ and highest at $V_{GS} = V_p$. The curves in the saturation region slope similarly to those of a bipolar transistor. These curves can likewise be extrapolated back to an intercept point analogous to the Early voltage, V_a, in the case of bipolar transistors. The slope of this line in bipolar transistors was attributed to the effects of basewidth modulation as V_{ce} in-

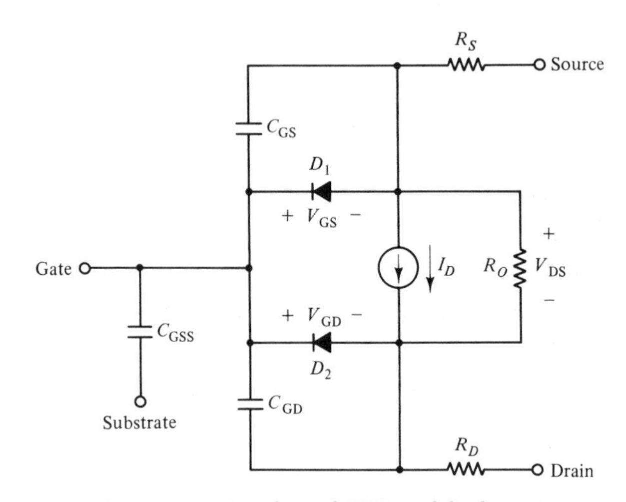

Figure 3-21 A p-channel JFET model schematic.

are not as obvious as others. The following is a list of conditions that can cause fault conditions to occur. They should be considered "red flags" while finalizing the system-level design, the design of a semicustom IC, or while investigating reliability questions or circuit failures.

- Signal applied with chip power off
- Signal present before chip power during system turn-on
- Failure or shutdown of one power supply on a dual supply
- Noise spikes on power supply lines from relays or motors
- Significant power supply fluctuations with reactive components connected to chip
- Reactive components connected to chip
- Signal swings that exceed positive or negative power supply
- Noise spikes on signal lines
- Electrostatic discharge during assembly, test or service

It is important to note that the occurrence of a damaging fault condition will not necessarily result in an instant failure. An overstressed junction, especially in an input stage, can cause performance degradation and later failure. Electrostatic discharge (ESD), once thought not to affect bipolar devices, can be a significant threat, especially on high-performance, small-geometry processes. ESD can cause immediate failure or a defect that can lead to failure in the longer term. Care should be taken to properly train all assembly and test personnel in handling procedures for ESD-sensitive devices. ESD precautions are extremely important in drier climates. Proper grounding should be used at all times. Even discharges that cannot be felt can be fatal to an IC. Insist that all devices be shipped from the vendor in antistatic packaging.

References

1. Gray, P. E., DeWitt, D., Boothroyd, A. R., and Gibbons, J. F., *Physical Electronics and Circuit Models of Transistors*. Wiley & Sons, New York, 1964; p. 20.
2. Gray *et al.*, (1964), Chapter 2.
3. Gray, P. R., Meyer, R. G., *Analysis and Design of Analog Integrated Circuits*. Wiley & Sons, New York, 1977; Chapter 1.
4. Grove, A. S., *Physics and Technology of Semiconductor Devices*. Wiley & Sons, New York, 1967; Chapter 6.
5. Gray and Meyer, (1977), p. 5–7.
6. Getreu, I., *Modeling the Bipolar Transistor*. Tektronix, Oregon, 1976; Chapter 2.

ac Analysis Invoked by the .AC card. This analysis performs the small-signal sinusoidal steady state transfer function of the circuit. This is usually voltage gain and phase response.

Noise Analysis Invoked by the .NOISE card and the .AC card (the ac analysis must be run if the noise analysis is to be run). This analysis calculates resistor thermal noise, semiconductor flicker, and shot noise. The thermal noise from bulk semiconductor material can be specified separately. Each noise source is assumed to be statistically independent and is calculated separately. The contributions of all noise sources are combined in an RMS fashion to determine the total noise at the specified output. This total noise is divided by the gain from the specified input to the specified output to calculate the equivalent input noise.

Distortion Analysis Invoked by the .DISTO card and the .AC card (the ac analysis must be run if the distortion analysis is to be run). Distortion analysis calculates both harmonic distortion and intermodulation distortion products of diodes and bipolar transistors as a power into a user-specified RLOAD. The distortion products are the powers in RLOAD of the second and third harmonics of the excitation frequency, F1. The intermodulation distortion products are the relative powers delivered to RLOAD of F1 + F2, F1 − F2, and 2F1 − F2, given the two excitation frequencies F1 and F2. The frequency F1 is controlled by the .AC card. The amplitude and frequency of F2 relative to F1 are specified in the .DISTO card.

Time-Domain Analysis

The time-domain analysis calculates the transient response of the circuit and can also perform a discrete fourier transform (DFT) on the output and calculate the total harmonic distortion of the circuit.

Transient Analysis Invoked by the .TRAN card and can be performed using either the results of the default bias-point calculation as the initial condition or by substituting user-selected initial conditions. To specify the initial conditions, the default bias-point calculation must be omitted with a special command (UIC—use initial conditions) in the .TRAN card. The desired initial conditions are then specified on the .IC card. The start time, timestep, and stop time are user-selectable.

Fourier Analysis Invoked by the .FOUR card. A DFT is performed to calculate the dc component and the amplitude and phase of the first nine frequency components. The percent total harmonic distortion is then calculated from these results. When using the Fourier transform, several precautions are necessary. One transient analysis and one Fourier analysis are performed during a run. The time window specified for the transient analysis will be the data record on which the DFT is performed. If the initial turn on of a circuit is being observed, be aware that the Fourier analysis will show frequency components (and distortion) associated with the turn on transients. If the Fourier analysis of the circuit in steady state is desired, be sure to delay the start of the transient analysis until the circuit has reached the steady state and all transient activity has subsided. Also, choose the time window such that the waveform being observed is continuous in the data record, since windowing the data prior to performing the DFT is not possible in SPICE.

4.2 SPICE Input Format

The SPICE input file, also referred to as a netlist, can be generated by optional schematic capture packages or by hand using an ASCII text editor. The orientation of this book is toward the manual entry of the SPICE input file. There are many different schematic capture packages running on a wide variety of computer platforms and it would not be practical to discuss them here. The format of the actual SPICE input file, however, is universal and whether generated by hand or software must conform to the standards that are about to be discussed.

General Input Requirements

The file must be created by an ASCII text editor or a word processor capable of storing the finished file in the ASCII (text) format. Characters used for formatting in non-ASCII word processor files can cause problems when included in a SPICE input file. The first card (line) in the input file *must* be the title of the circuit or file. If this card is omitted or left blank, the first card in the file will be assumed by SPICE to be the title card and the run will bomb. The last card (line) in the file *must* be an .END card. This is how SPICE can tell that the file has ended. If this card is omitted, an error message will result. All SPICE input, except for comments, *must* use capital letters. Any card with an asterisk in the first column will be considered a comment card.[1] It will be printed in the file listing but will be ignored otherwise. Placing an asterisk in the first column of any card allows that card to be edited out as far as SPICE execution is concerned without actually removing it. Each line in the input file can contain up to 72 characters. Any input after 72 characters will be truncated and no error message will result. Some control statements such as .MODEL or a piecewise linear description frequently require more than 72 characters. This can be accommodated with the use of a continuation card. Any card (line) that has a plus sign in the first column will be considered by SPICE as a continuation of the previous line.[2] Data fields (times, frequencies, amplitudes, model parameters, interconnection nodes, etc.) can be separated by spaces, commas, equal signs, or parentheses. Optional parameters are either assumed to be zero or are set to a default value unless defined in the input file.

Circuit Description

The circuit connection is defined for SPICE with node numbers. A node is any point in the circuit with at least two connections and is identified with a positive integer between 0 and 999 (higher node numbers are possible with certain restrictions.[3] See the user's guide for your particular version of SPICE). Node 0 is universally reserved as ground. Node 0 must be used as the ground node in a subcircuit description (schematic) but must never be used as a connection node in the .SUBCKT card. If a dc analysis of the circuit is desired, there must be a dc path to ground (which may be a very high value resistor) at each node.[4]

Component Names

All components in a circuit must have unique names so that SPICE can distinguish between them. Component names are the first continuous block of characters on the component definition line. A component name must begin with the key letter for the desired component and thereafter can contain up to seven additional letters and numbers in any order. The key letters are as follows:

Passive components
- C = Capacitor
- K = Coupled inductor
- L = Inductor
- R = Resistor
- T = Transmission line

Sources
- E = Voltage-controlled voltage source
- F = Current-controlled current source
- G = Voltage-controlled current source
- H = Current-controlled voltage source
- I = Independent current source
- V = Independent voltage source

Semiconductors
- D = Diode
- J = Junction field-effect transistor (JFET)
- M = Metal oxide semiconductor field-effect transistor (MOSFET)
- Q = Bipolar transistor

Subcircuit
- X = Subcircuit

The component name can be of any length but only the first eight characters, including the component designation letter, are recognized by SPICE. For example,

CAPACITOR1 = CAPACITOR2 = CAPACITO

It is best to use numbered component names such as R12, C25, or Q10 for most components and reserve the more textual names for special components such as RLOAD, RINPUT, CLOAD, QOUTPUT, and QDRIVER.

Number Scaling

SPICE accepts numerical inputs in several different formats. Numbers can be entered directly as integers or real numbers. They can also be entered in scientific notation as an integer or real value with a power of 10 (i.e., 1.76E-3). Alternatively, letters can be used to scale values in the following commonly used engineering units:[2]

Symbol	Name	Scale factor
F	femto	1E-15
P	pico	1E-12
N	nano	1E-9
U	micro	1E-6
M	milli	1E-3
K	kilo	1E3
MEG	mega	1E6
G	giga	1E9
T	tera	1E12

These engineering unit scale factors are very convenient and make the input easier to read. There are, however, several important points to note about their use. Only one scale factor can be used at a time. Any letters after the first recognized scale factor are ignored by SPICE. The following examples illustrate these points:[2]

1KV = 1K = 1E3 = 1000 (The V is ignored)
1KKOHM = 1K = 1E3 = 1000 (KOHM is ignored)
1UUF = 1U = 1E-6 = .000001 (UF is ignored)

Output

Every SPICE run, in order to be meaningful, must have some sort of output specified. In standard SPICE there are two types of output available: .PRINT[5] and .PLOT.[6] The .PRINT command provides a tabular listing of the specified node voltages. The .PLOT command provides a line printer-type pseudographical plot of the specified node voltages. Most graphic post processors require .PRINT output.

SPICE Input File Format

The following is a suggested format for a SPICE input file. By using this as a template in your text editor and "filling in the blanks" when a new input file is created, many input errors can be avoided.

```
CIRCUIT NAME:                    BY:
*CIRCUIT DESCRIPTION:
*
******CIRCUIT CONNECTION******
***RESISTORS***


***CAPACITORS***


***INDUCTORS***


***SEMICONDUCTORS***


***SUBCIRCUITS***


***SOURCES***


******MODEL DEFINITION******


******SUBCIRCUIT DEFINITION******


******ANALYSIS CONTROL******


******OUTPUT CONTROL******


.END
```

4.3 Passive Components

This section will describe in detail the definition statements for each type of passive component. In each case, there is a minimum amount of information that must be given for SPICE to properly identify the component, connect it in the circuit, and assign the proper value to it. Optional parameters can also be added to make the component more accurately model the behavior of the real component.

Resistors

The definition statement for a resistor contains four mandatory data fields and three optional data fields separated by spaces. The format for a resistor definition statement is as follows:

```
RXXXXXXX   N1   N2   VALUE   TC = TC1,TC2
```

The resistor definition statement must begin with the key letter R. The R can be followed by up to seven additional letters or numbers to uniquely identify the resistor. Next are the node1 and node2 fields. These fields describe the node numbers in the circuit to which the resistor is connected. The value field defines the resistor value in ohms and can be positive or negative but cannot be zero. This value is the resistor value at the nominal temperature TNOM. The default value for TNOM is 27°C but can be changed to any desired value on the .OPTIONS control card. The optional temperature coefficients TC1 and TC2 can be added to cause the resistor value to change with temperature. They can be positive or negative and are assumed to be zero if unspecified. The equation used by SPICE to describe the resistor value over temperature is[7,8]

$$R(T) = R(TNOM)(1 + TC1(T - TNOM) + TC2(T - TNOM)^2)$$

Capacitors

The definition statement for a capacitor contains four mandatory data fields and an optional data field separated by spaces. The format for a capacitor definition statement is as follows:

```
CXXXXXXX   N+   N-   VALUE   IC = INCOND
```

The capacitor definition statement must begin with the key letter C. The C can be followed by up to seven additional letters or numbers to uniquely identify the capacitor. Next are the node+ and node− fields. Capacitor nodes are polarized so the polarity of initial conditions can be determined. These fields describe the node numbers in the circuit to which the capacitor is connected. The value field defines the capacitor value in farads. The optional initial condition field, IC = INCOND, allows the time-zero capacitor voltage (in volts), INCOND, to be set. This initial condition applies only if the UIC option is selected on the .TRAN control card.

Nonlinear capacitors can be described as follows:

CXXXXXXX N+ N− POLY C0 C1 C2 C3 . . . IC = INCOND

where the value of the capacitor is determined by the polynomial

$$\text{VALUE} = \text{C0} + \text{C1V} + \text{C2V}^2 + \text{C3V}^3 + \ldots$$

The key word POLY tells SPICE that the value of the capacitor will be determined by a polynomial and that the next series of data fields, separated by spaces, are the polynomial's coefficients. As before, IC = INCOND is an optional data field that describes the initial charge on the capacitor at time zero and applies only if the UIC option is selected on the .TRAN control card.[9,10]

Inductors

The definition statement for an inductor contains four mandatory data fields and an optional data field separated by spaces. The format for an inductor definition statement is as follows:

LXXXXXXX N+ N− VALUE IC = INCOND

The inductor definition statement must begin with the key letter L. The L can be followed by up to seven additional letters or numbers to uniquely identify the inductor. Next are the node+ and node− fields. Inductor nodes are polarized so the polarity of initial conditions can be determined. These fields describe the node numbers in the circuit to which the inductor is connected. The value field defines the inductor value in henries. The optional initial condition field, IC = INCOND, allows the time-zero inductor current (in amps), INCOND, to be set. This initial condition applies only if the UIC option is selected on the .TRAN control card.

Nonlinear inductors can be described as follows:

LXXXXXXX N+ N− POLY L0 L1 L2 L3 . . . IC = INCOND

where the value of the inductor is determined by the polynomial

$$VALUE = L0 + L1I + L2I^2 + L3I^3 + \ldots$$

The key word POLY tells SPICE that the value of the inductor will be determined by a polynomial and that the next series of data fields, separated by spaces, are the polynomial's coefficients. As before, INCOND is the initial current through the inductor at time zero and applies only if the UIC option is selected on the .TRAN control card.[9,11]

Coupled Inductors

The definition statement for a coupled inductor contains four mandatory data fields separated by spaces. The format for a coupled inductor definition statement is as follows:

KXXXXXXX LYYYYYYY LZZZZZZZ VALUE

The coupled inductor definition statement must begin with the key letter K. The K can be followed by up to seven additional letters or numbers to uniquely identify the coupled inductor. LYYYYYYY and LZZZZZZZ are the two coupled inductors. The value field defines the coupling coefficient, which must be greater than 0 and less than or equal to 1. Place the "dot" on the first node (N+) of each inductor to determine the relative phasing of the two inductors.[12,13]

Transmission Lines (Lossless)

The definition statement for a transmission line contains seven mandatory data fields and three optional data fields separated by spaces. The format for a transmission line definition statement is as follows:

```
TXXXXXXX  N1  N2  N3  N4  ZO = VALUE
+ TD = VALUE F = FREQ  NL = NRMLEN
+ IC = V1,I1,V2,I2
```

The transmission-line definition statement must begin with the key letter T. The T can be followed by up to seven additional letters or numbers to uniquely identify the transmission line. Next are the node fields. N1 and N2 are the nodes at port 1 and N3 and N4 are the nodes at port 2. These fields describe the node numbers in the circuit to which the resistor is connected. The next field, ZO, is the characteristic impedance of the transmission line. The last mandatory data is the time delay. TD can be specified directly (i.e., TD = 25NS) or TD can be omitted and a frequency, F, and the normalized electrical length, NL, of the line at F can be specified and SPICE will calculate TD. NL defaults to 0.25 (quarter wavelength) if F is given and NL is omitted. One of these methods must be used. The optional initial condition specification sets the voltage and current at both transmission line ports at time zero. These conditions apply only if the UIC option is selected on the .TRAN card.[14,15]

4.4 Current and Voltage Sources

SPICE offers four types of voltage and current sources: linear dependent sources, nonlinear dependent sources, time-independent independent sources, and time-dependent independent sources. The value of a dependent source is determined by its gain and a current or voltage present in the circuit. (Note: the voltage of a current controlled voltage source can be controlled by the current flowing through it, thus modeling a resistor.) An independent source (either time-independent or time-dependent) is a forcing function with a value independent of circuit function. In all cases, current is assumed to flow from the positive source terminal (N+) through the source to the negative source terminal (N−). The current for a controlling voltage source is likewise assumed to flow from the positive controlling source terminal (NC+) through the source to the negative controlling source terminal (NC−). Sources do not have to be connected to ground. A 0-V independent voltage source can serve as an ammeter or as a controlling source when the current in a connecting "wire" of the circuit is needed to control a dependent source. The insertion of a 0-V source anywhere in a circuit will have no effect on circuit performance since SPICE will view it as a short circuit.

Linear Dependent Sources

Linear dependent sources are characterized by the following equations:[16]

$I = G^*V$	Voltage-controlled current source
$V = E^*V$	Voltage-controlled voltage source
$I = F^*I$	Current-controlled current source
$V = H^*I$	Current-controlled voltage source

where

G = transconductance
E = voltage gain
F = current gain
H = transresistance

Linear Voltage-Controlled Current Sources

The definition statement for a linear voltage-controlled current source contains six mandatory data fields separated by spaces. The definition statement format is as follows:

```
GXXXXXXX  N+   N-   NC+   NC-   G
```

The linear voltage-controlled current source definition statement must begin with the key letter G. The G can be followed by up to seven additional characters to uniquely identify the source. Next are the N+ and N− fields. These fields describe the node numbers in the circuit to which the positive (+) and negative (−) terminals of the source are connected. The NC+ and NC− fields describe the the positive (+) and negative (−) terminals of the controlling source. The "G" field is the transconductance value in mhos.[17,18]

Examples

```
G5   7   0   0   1.33MMHO
GIN   4   0   11   0   1.73E-6
```

Linear Voltage-Controlled Voltage Sources

The definition statement for a linear voltage-controlled voltage source contains six mandatory data fields separated by spaces. The definition statement format is as follows:

```
EXXXXXXX  N+  N-  NC+  NC-  E
```

The linear voltage-controlled voltage source definition statement must begin with the key letter E. The E can be followed by up to seven additional characters to uniquely identify the source. Next are the N+ and N− fields. These fields describe the node numbers in the circuit to which the positive (+) and negative (−) terminals of the source are connected. The NC+ and NC− fields describe the the positive (+) and negative (−) terminals of the controlling source. The "E" field is the voltage gain.[19,20]

Examples

```
EOUT  5  7  11  33  23.2
E1  4  0  2  0  5
```

Linear Current-Controlled Current Sources

The definition statement for a linear current-controlled current source contains five mandatory data fields separated by spaces. The definition statement format is as follows:

```
FXXXXXXX  N+  N-  VNAME  F
```

The linear current-controlled current source definition statement must begin with the key letter F. The F can be followed by up to seven additional characters to uniquely identify the source. Next are the N+ and N− fields. These fields describe the node numbers in the circuit to which the positive (+) and negative (−) terminals of the source are connected. The VNAME field describes the name of the voltage source (which serves as an ammeter) through which the control current flows. The "F" field is the current gain.[21,22]

Examples

```
FOUT  3  0  EIN  3E2
F2  12  0  ECONTROL  7.8
```

Linear Current-Controlled Voltage Sources

The definition statement for a linear current-controlled voltage source contains five mandatory data fields separated by spaces. The definition statement format is as follows:

HXXXXXXX N+ N− VNAME H

The linear current-controlled voltage source definition statement must begin with the key letter H. The H can be followed by up to seven additional characters to uniquely identify the source. Next are the N+ and N− fields. These fields describe the node numbers in the circuit to which the positive (+) and negative (−) terminals of the source are connected. The VNAME field describes the name of the voltage source (which serves as an ammeter) through which the control current flows. The "H" field is the transresistance in ohms.[23,24]

Examples

H1	23	5	VSENSE	1E3
H17	4	0	VTRAN	3K

Nonlinear Dependent Sources

Nonlinear dependent sources are characterized by the following equations:[25]

$I = F(V)$	Voltage-controlled current source
$V = F(V)$	Voltage-controlled voltage source
$I = F(I)$	Current-controlled current source
$V = F(I)$	Current-controlled voltage source

The function of voltage or current is a polynomial with an arbitrary number of dimensions and coefficients that describe transconductance, voltage gain, current gain, and transresistance. The polynomial is described by coefficients P0, P1, P2, . . . , PN. If the polynomial has one dimension and one coefficient is given, SPICE assumes the given coefficient is P1 and P0 = 0. The NC+ and NC− fields describe the the positive (+) and negative (−) terminals of the controlling sources. One pair of control nodes (node numbers are separated by a space) must be included for each dimension. In other words, the number of control node pairs must equal ND (number of dimensions). If an optional initial condition parameter(s) is not specified, SPICE assumes a value of 0 for the controlling variable(s). After initial convergence is obtained, SPICE will continue to iterate to the exact solution. An accurate "guess" of the true initial value of all controlling variable(s) will reduce computation time.

Nonlinear Voltage-Controlled Current Source

The definition statement format for a nonlinear voltage-controlled current source is as follows:

```
GXXXXXXX   N+      N-      POLY(ND)      NC1+      NC1-
+...... NCN+     NCN-      P0     P1      ..... PN
+ IC = IC1,. . .,ICN
```

The nonlinear voltage-controlled current source definition statement must begin with the key letter G. The G can be followed by up to seven additional characters to uniquely identify the source. Next are the N+ and N− fields. These fields describe the node numbers in the circuit to which the positive (+) and negative (−) terminals of the source are connected. The transductance of the source is described by a polynomial. POLY(ND) (ND = number of dimensions for the polynomial) must be included only if ND is greater than the default value of one.[18]

Nonlinear Voltage-Controlled Voltage Source

The definition statement format for a nonlinear voltage-controlled voltage source is as follows:

```
EXXXXXXX   N+      N-      POLY(ND)      NC1+      NC1-
+ ...... NCN+     NCN-      P0     P1      ...... PN
+ IC = IC1, . . . ICN
```

The nonlinear voltage-controlled voltage source definition statement must begin with the key letter E. The E can be followed by up to seven additional characters to uniquely identify the source. Next are the N+ and N− fields. These fields describe the node numbers in the circuit to which the positive (+) and negative (−) terminals of the source are connected. The gain of the source is described by a polynomial. POLY(ND) (ND = number of dimensions for the polynomial) must be included only if ND is greater than the default value of one.[20]

Nonlinear Current-Controlled Current Source

The definition statement format for a nonlinear current-controlled current source is as follows:

```
FXXXXXXX  N+    N-    POLY(ND)    NC1+    NC1-
+ ....... NCN+    NCN-    P0    P1    ....... PN
+ IC = IC1,. . .,ICN
```

The nonlinear current-controlled current source definition statement must begin with the key letter F. The F can be followed by up to seven additional characters to uniquely identify the source. Next are the N+ and N− fields. These fields describe the node numbers in the circuit to which the positive (+) and negative (-) terminals of the source are connected. The current gain of the source is described by a polynomial. POLY(ND) (ND = number of dimensions for the polynomial) must be included only if ND is greater than the default value of one.[22]

Nonlinear Current-Controlled Voltage Source

The definition statement format for a nonlinear current-controlled voltage source is as follows:

```
HXXXXXXX  N+    N-    POLY(ND)    NC1+    NC1-
+ ....... NCN+    NCN-    P0    P1    ....... PN
+ IC = IC1,. . .,ICN
```

The nonlinear current-controlled voltage source definition statement must begin with the key letter H. The H can be followed by up to seven additional characters to uniquely identify the source. Next are the N+ and N− fields. These fields describe the node numbers in the circuit to which the positive (+) and negative (−) terminals of the source are connected. The transresistance of the source is described by a polynomial. POLY(ND) (ND = number of dimensions for the polynomial) must be included only if ND is greater than the default value of one.[24]

Examples of polynomials One-dimensional (one controlling source, S1, with controlling nodes NC1+, NC1− and P0, P1, P2, P3, P4, and P5 specified):

```
G1DIM  5    7    POLY(1)    2    3
+ 1, 3, .47, .86, .5
```

The polynomial that describes G1DIM is

$$P_0 + P_1S_1 + P_2S_1^2 + P_3S_1^3 + P_4S_1^4 + P_5S_1^5$$

Two-dimensional (two controlling sources, S1 and S2, with controlling nodes NC1 +, NC1 −, NC2 +, NC2 − and P0 . . . P9 specified):

```
G2DIM   5    7    POLY(2)    2    3    5    8
+ 1, 1.2, .44, .56, 1.7, .6, 1.8, 1.1, .2
```

The polynomial that describes G2DIM is

$$P_0 + P_1S_1 + P_2S_2 + P_3S_1^2 + P_4S_1S_2 + P_5S_2^2 + P_6S_1^3 + P_7S_1^2S_2 + P_8S_1S_2^2 + P_9S_2^3$$

Three-dimensional (three controlling sources, S1, S2, and S3, with controlling nodes NC1 +, NC1 −, NC2 +, NC2 −, NC3 +, NC3 −, and P0 . . . P20 specified):

```
G3DIM   5    7    POLY(3)    2    3    6    5    9    1
+ 5, 4, 1.3, .4, 1, .5, 1.6, .3, 1.8, 1.3, .22, .4, .7, .8, 1.9,
+ 1.3, 1, .7, .9, .022
```

The polynomial that describes G3DIM is

$$P_0 + P_1S_1 + P_2S_2 + P_3S_3 + P_4S_1^2 + P_5S_1S_2 + P_6S_1S_3 + P_7S_2^2 + P_8S_2S_3 + P_9S_3^2 + P_{10}S_1^3 + P_{11}S_1^2S_2 + P_{12}S_1^2S_3 + P_{13}S_1S_2^2 + P_{14}S_1S_2S_3 + P_{15}S_1S_3^2 + P_{16}S_2^3 + P_{17}S_2^2S_3 + P_{18}S_2S_3^2 + P_{19}S_3^3 + P_{20}S_1^4$$

Independent Sources

Time-Independent Independent Sources

Time-independent independent sources are typically used for power supplies, fixed bias, reference voltages, ammeters, or small-signal ac analysis.

dc Sources

The definition statement for a dc time-independent independent source has four mandatory fields separated by spaces (the “DC” field is optional). The definition statement format is as follows:

VXXXXXXX N+ N− DC VALUE

for a voltage source or

IXXXXXXX N+ N− DC VALUE

for a current source. Note: "DC" can be omitted from the voltage source definition statement if the source is used as a constant source such as a power supply as long as "VALUE" is included. "DC" and "VALUE" can be omitted if the source is being used as a 0-V ammeter.

The voltage and current source definition statement must begin with the key letters V and I, respectively. The V or I can be followed by up to seven additional characters to uniquely identify the source. Next are the N+ and N− fields. These fields describe the node numbers in the circuit to which the positive (+) and negative (−) terminals of the source are connected. Positive current flows from the positive node through the source to the negative node for positive-valued voltage and current sources. "DC" specifies a constant voltage or current output from these sources from time zero. The dc value of a source is also used as the source value for transient analysis. If the source value is omitted, a value of 0 will be assumed for transient analysis. Voltage sources need not be grounded and can be inserted anywhere in a circuit. Current sources, on the other hand, must have one terminal connected to ground. "VALUE" is the value of the source in volts or amperes. If the value is negative, current flow is reversed.

ac Sources

The definition statement for an ac time-independent independent source has four mandatory fields and three optional fields. The definition statement format is as follows:

VXXXXXXX N+ N− DCVALUE AC ACMAG
+ ACPHASE

for a voltage source or

IXXXXXXX N+ N− DCVALUE AC ACMAG
+ ACPHASE

for a current source.

The voltage and current source definition statement must begin with the key letters V and I, respectively. The V or I can be followed by up to seven additional characters to uniquely identify the source. Next are the N+ and N− fields. These fields describe the node numbers in the circuit to which the positive (+) and negative (−) terminals of the source are connected. Positive current flows from the positive node through the source to the negative node for positive valued voltage and current sources. "DCVALUE" specifies a constant voltage or current output from these sources at time zero and is used as the source value for transient analysis. If "DCVALUE" is omitted, a value of 0 will be assumed for transient analysis. The key word "AC" is included only if the source is an input for small-signal ac analysis. "ACMAG" and "ACPHASE" are the relative amplitude and phase of the source during this analysis. If the key word "AC" is present and "ACMAG" and "ACPHASE" are omitted, SPICE assumes an amplitude of 1 and a phase of 0. Sources need not be grounded and can be inserted anywhere in a circuit.

Time-Dependent Independent Sources

An independent source can be assigned a time-dependent value for transient analysis. The time-zero value of the source is used to calculate the dc bias point. The time-dependent value can be one of the following functions: pulse, exponential, sinusoidal, piece-wise linear, and single-frequency FM.

The definition statement for a time-dependent independent source has four mandatory fields separated by spaces. The definition statement format is as follows:

```
VXXXXXXX   N+   N-   WAVEFORM
```

for a voltage source or

```
IXXXXXXX   N+   N-   WAVEFORM
```

for a current source.

The voltage and current source definition statement must begin with the key letters V and I, respectively. The V or I can be followed by up to seven additional characters to uniquely identify the source. Next are the N+ and N− fields. These fields describe the node numbers in the circuit to which the positive (+) and negative (−) terminals of the source are connected. Positive current flows from the positive node through the source to the negative node for positive-valued voltage and current sources. Sources need not be grounded and can be inserted anywhere in a circuit. The "WAVEFORM" field is a description of the time-dependent waveform and will have one of the following forms: pulse, sinusoidal, exponential, piece-wise linear, or single-frequency FM.

Pulse

PULSE(V1 V2 TD TR TF PW PER)

The fields within the parentheses describe the characteristics of the waveform. V1 and V2 are the initial and pulsed value of the waveform in volts or amperes and must be specified. The remaining optional fields describe the time characteristics of the waveform in seconds. If these fields are omitted, the default values listed below will be assumed by SPICE where TSTEP, the printing increment, and TSTOP, the final time, are specified on the .TRAN card.

Field	Default value
TD (delay time)	0
TR (rise time)	TSTEP
TF (fall time)	TSTEP
PW (pulse width)	TSTOP
PER (period)	TSTOP

Examples

```
VPULSE  1   0   PULSE(0  5  10N  4N  4N  16N  40N)
V1  3   7   PULSE(0  10)
```

Sinusoidal

SIN(VO VA FREQ TD THETA)

The fields within the parentheses describe the characteristics of the waveform. VO and VA are the offset and amplitude of the waveform in volts or amperes and must be specified. The remaining optional fields describe the time characteristics of the waveform. FREQ is specified in hertz, TD in seconds, and THETA, the damping factor, in 1/sec. If these fields are omitted, the default values listed below will be assumed by SPICE where TSTOP, the final time, is specified on the .TRAN card.

Field	Default value
FREQ (frequency)	1/TSTOP
TD (delay time)	0
THETA (damping factor)	0

The amplitude of the waveform is VO from time zero to TD. From TD to TSTOP the amplitude is described by the following equation:

$$\text{Amplitude} = \text{VO} + \text{VA}\exp[-(\text{TIME} - \text{TD})\text{THETA}]\\ \sin[2\pi\text{FREQ}(\text{TIME} + \text{TD})]$$

Examples

```
V1  4  5  SIN(0  5  1MEG  3N  0)
V2  3  2  SIN(5  7)
```

Exponential

```
EXP(V1  V2  TD1  TAU1  TD2  TAU2)
```

The fields within the parentheses describe the characteristics of the waveform. V1 and V2 are the initial and pulsed value of the waveform in volts or amperes and must be specified. The remaining optional fields describe the time characteristics of the waveform and have units of seconds. TD1 is the rise delay time, TAU1 is the rise time constant, TD2 is the fall delay time, and TAU2 is the fall time constant. If these fields are omitted, the default values listed below will be assumed by SPICE where TSTEP, the time increment used for printing, is specified on the .TRAN card.

Field	Default value
TD1 (rise delay time)	0
Tau1 (rise time constant)	TSTEP
TD2 (fall delay time)	TD1 + TSTEP
TAU2 (fall time constant)	TSTEP

The amplitude of the waveform is V1 from time zero to TD1. From TD1 to TD2 the amplitude is described by the following equation:

$$\text{Amplitude} = V1 + (V2 - V1)\left(1 - \exp\left(\frac{-(\text{TIME} - \text{TD1})}{\text{TAU1}}\right)\right)$$

From TD2 to TSTOP the amplitude is described by the following equation:

$$\text{Amplitude} = V1 + (V2 - V1)\left(1 - \exp\left(\frac{-(\text{TIME} - \text{TD1})}{\text{TAU1}}\right)\right) + (V1 - V2)\left(1 - \exp\left(\frac{-(\text{TIME} - \text{TD2})}{\text{TAU2}}\right)\right)$$

Examples

```
VIN   0   9   EXP(0  5  10N  25N  10N  20N)
V2    0   1   EXP(0  5)
```

Piecewise Linear

PWL(T1 V1 T2 V2 . . . TN VN)

Each pair of parameters specifies the source value in volts or amperes, VX, at time TX from time zero. The value of the source at intermediate times is determined by linear interpolation.

Examples

```
VIN   0   9   PWL(10N  0  20N
                 1  25N  2  30N  2.5)
VSOURCE   0   7   PWL(1N  0.5  10N
                   1  100N  3.5  200N  5.1)
```

Single-Frequency FM

SFFM(VO VA FC MDI FS)

The fields within the parentheses describe the characteristics of the waveform. VO and VA are the offset and amplitude of the waveform in volts or amperes and must be specified. The remaining optional fields describe the time characteristics of the waveform. FC, the carrier frequency, is specified in hertz. The modulation index, MDI, is a number between 0 and 100, representing the percent modulation of the carrier by the signal. FS, the signal frequency, is specified in hertz. If these fields are omitted, the default values listed below will be assumed by SPICE where TSTOP, the final time, is specified on the .TRAN card.

Field	Default value
FC (carrier frequency)	1/TSTOP
MDI (modulation index)	0
FS (signal frequency)	1/TSTOP

The amplitude of the waveform is described by the following equation:

$$\text{Amplitude} = \text{V0} + \text{VA} \sin((2\pi \text{ FC TIME}) + \text{MDI} \sin(2\pi \text{ FS TIME}))$$

Examples

```
VIN      0    10    SFFM(0  1  1MEG  50  1K)
VSIGNAL   0   8    SFFM(2.5  1M  7.23MEG  80  1.73K)
```

For more information on independent sources see References 26 and 27.

4.5 Active Components

This section describes the use of diodes, bipolar junction transistors (BJTs), junction field-effect transistors (JFETs), and metal oxide semiconductor field effect transistors (MOSFETs). The use of the .MODEL card is described. The component definition statement is presented for each device, followed by a detailed description of the model and model parameters for that device. Examples are given for both the component description statement and .MODEL cards.

General Statements

A great many active components are typically used in an integrated design. SPICE requires a minimum number of device parameters in the component definition statement and allows semiconductor devices to reference a .MODEL card via a model name (MNAME) which completely defines the device. This technique avoids the unnecessary duplication of device parameters in the component definition statement. A model name, MNAME, and a corresponding .MODEL card can be assigned to each device type or geometry and referenced in the component definition statement of as many components as necessary. This approach is valuable when working with semicustom arrays since, out of the hundreds of available transistors, there are very few different devices. Another convenient feature of SPICE is the area factor. This parameter describes a diode, BJT, or JFET as being equivalent to the area factor number of paralleled devices of a specified model type. The model parameters ratioed by the area factor are indicated when the models are discussed. Frequently, in semicustom designs, several transistors are paralleled to create ratioed current sources. The area factor allows different geometry (size) devices to be specified in terms of each other. The default value for "AREA" is 1.

When circuits such as flip-flops that have more than one stable state are being simulated, the optional "OFF" specification can be used to ensure that the circuit starts in the proper state. The terminal voltages of a semiconductor device specified as "OFF" are set at zero for the initial dc operating point. SPICE will compute the exact operating point after initial convergence is achieved even if the "wrong" state was specified by the "OFF" specification. The .NODESET card can be used to achieve the same results by specifying initial node voltages in the circuit. The .NODESET card is generally more efficient than the "OFF" specification. Initial conditions for transient analysis are specified with the .IC card.

.MODEL Card

The general format of the .MODEL card is

```
.MODEL MNAME TYPE PNAME1 = PVALUE1 . . .
+PNAMEN = PVALUEN
```

The .MODEL card must begin with ".MODEL." Next, separated by a space, is the model name "MNAME." MNAME must begin with a letter but can consist of up to seven letters and numbers to uniquely name a model. The "TYPE" field identifies the model as one of the following seven device types that can be modeled using a .MODEL card:

TYPE	Description
NPN	*npn* BJT
PNP	*pnp* BJT
D	Diode
NJF	*n*-channel JFET
PJF	*p*-channel JFET
NMOS	*n*-channel MOSFET
PMOS	*p*-channel MOSFET

The "TYPE" identifier is followed by parameter value definitions for the model of the device "TYPE" just declared. Parameter values for a particular model are specified by equating a parameter name, PNAME1, to a corresponding parameter value, PVALUE1. A PNAME = PVALUE statement is separated from the next PNAME = PVALUE statement by a space. Model parameters are assigned their default value unless a different value is specified. Examples of .MODEL cards for each device are given after the model for the device is presented.[28,29]

Junction Diode

The definition statement for a junction diode contains four mandatory data fields and three optional data fields separated by spaces. The format for a diode definition statement is as follows:

DXXXXXXX N+ N− MNAME AREA OFF IC = VD

The diode definition statement must begin with the key letter D. The D can be followed by up to seven additional letters or numbers to uniquely identify the diode. Next are the N+ and N− fields. These fields describe the node numbers in the circuit to which the positive (anode) and negative (cathode) terminals of the diode are connected. "MNAME" is the name of the model describing the diode. The optional "AREA" is the area ratio of the diode being defined to that of the model. The optional "OFF," if included, sets the initial terminal voltage of the diode to zero. "IC = VD" is used in conjunction with the UIC command on the .TRAN card to set the initial voltage of the diode for transient analysis at a voltage other than that determined by the dc bias-point calculation.[30,31]

Diode Model

The following is a list of the diode model parameters and default values:[31,32]

Parameter	Units	Default value
IS (saturation current)	A	1.0E-14
RS (ohmic resistance)	Ω	0
N (emission coefficient)		1
TT (transit time)	sec	0
CJO (zero-bias junction capacitance)	F	0
VJ (junction potential)	V	1
M (grading coefficient)		0.5
EG (activation energy)	eV	1.11
XTI (temperature exponent of IS)		3.0
KF (flicker noise coefficient)		0
AF (flicker noise exponent)		1
FC (coeff. for forward-biased capacitance)		0.5
BV (reverse breakdown voltage)	V	Infinite
IBV (current at BV)	A	1.0E-3

Examples

```
.MODEL    DIODE1    D    IS=1.7E-15
+ RS=2    CJO=4PF    VJ=.79 + BV=50    IBV=200E-6

.MODEL    DIODE2    D    VJ=0.65    BV=20V    CJO=1PF
+ IBV=10E-6
```

Bipolar Transistors

The definition statement for a bipolar junction transistor (BJT) contains five mandatory data fields and four optional data fields separated by spaces. The format for a BJT definition statement is as follows:

```
QXXXXXXX    NC    NB    NE    NS    MNAME    AREA
+ OFF    IC = VBE, VCE
```

The BJT definition statement must begin with the key letter Q. The Q can be followed by up to seven additional letters or numbers to uniquely identify the BJT. Next are the NC, NB, and NE fields. These fields describe the node numbers in the circuit to which the collector, base, and emitter of the transistor are connected. The substrate node, NS, is optional and defaults to ground (not the negative supply) if not specified. "MNAME" is the name of the model describing the BJT. The optional "AREA" is the area ratio of the BJT being defined to that of the model. The optional "OFF," if included, sets the initial terminal voltages of the BJT to zero. "IC = VBE, VCE" is used in conjunction with the UIC command on the .TRAN card to set the initial terminal voltages of the diode for transient analysis to a voltage other than that determined by the dc bias-point calculation.[33,34,35]

Bipolar Junction Transistor Model

The model used by SPICE is based on the Gummel–Poon model when sufficient model parameters are specified. Otherwise, the model reverts to the simpler Ebers–Moll model. The following is a list of the BJT model parameters and default values:[36]

Parameter	Units	Default value
IS (saturation current)	A	1.0E−16
BF (max. ideal forward beta)		100
NF (forward emission coefficient)		1.0
VAF (forward Early voltage)	V	Infinite
IKF (current to start forward beta high-current roll-off)	A	Infinite
ISE (B-E leakage saturation current)	A	0
NE (B-E leakage emission coefficient)		1.5
BR (max. ideal reverse beta)		1.0
NR (reverse emission coefficient)		1.0
VAR (reverse Early voltage)	V	Infinite
IKR (current to start reverse beta high-current roll-off)	A	Infinite
ISC (B-C leakage saturation current)	A	0
NC (B-C leakage emission coefficient)		2.0
RB (zero-bias base resistance)	Ω	0
IRB (current for base resistance = half of minimum value)	A	Infinite
RBM (min. base resistance at high currents)	Ω	RB

RE (emitter resistance)	Ω	0
RC (collector resistance)	Ω	0
CJE (B-E zero-bias depletion capacitance)	F	0
VJE (B-E built-in potential)	V	0.75
MJE (B-E junction potential factor)		0.33
TF (ideal forward transit time)	sec	0
XTF (bias dependence coefficient of TF)		0
VTF (V_{bc} dependence of TF)	V	Infinite
ITF (high-current dependence of TF)	A	0
PTF (excess phase at $f = 1/(2\pi TF)$ Hz)	deg	0
CJC (B-C zero-bias depletion capacitance)	F	0
VJC (B-C built-in potential)	V	0.75
MJC (B-C junction exponential factor)		0.33
XCJC (fraction of CJC connected to base)		1.0
TR (ideal reverse transit time)	sec	0
CJS (zero-bias collector-substrate C)	F	0
VJS (epi-substrate built-in potential)	V	0.75
MJS (epi-substrate exponential factor)		0
XTB (forward and reverse beta temperature exponent)		0
EG (energy gap for T dependence of IS)	eV	1.11
XTI (temperature exponent for IS)		3.0
KF (flicker-noise coefficient)		0
AF (flicker-noise exponent)		1.0
FC (forward-bias depletion capacitance coefficient)		0.5

Examples

```
.MODEL    NPN1      NPN     BF=50
.MODEL    QN3904    NPN     BF=300    VAF=325
```

Junction Field-Effect Transistors

The definition statement for a junction field effect transistor (JFET) contains five mandatory data fields and three optional data fields separated by spaces. The format for a JFET definition statement is as follows:

```
JXXXXXXX   ND   NG   NS   MNAME   AREA   OFF
+ IC = VDS,VGS
```

The JFET definition statement must begin with the key letter J. The J can be followed by up to seven additional letters or numbers to uniquely identify the JFET. Next are the ND, NG, and NS fields. These fields describe the node numbers in the circuit to which the drain, gate, and source of the transistor are connected. "MNAME" is the name of the model describing the JFET. The optional "AREA" is the area ratio of the JFET being defined to that of the model. The optional "OFF," if included, sets the initial terminal voltages of the JFET to zero. "IC = VDS, VGS" is used in conjunction with the UIC command on the .TRAN card to set the initial terminal voltages for transient analysis to a voltage other than that determined by the dc bias-point calculation.[37,38]

Junction Field-Effect Transistor Model

The model used by SPICE is based on the Shichman–Hodges model. The following is a list of the JFET model parameters and default values:[37,39]

Parameter	Units	Default value
VTO (pinch-off voltage)	V	−2.0
BETA (transconductance)	A/V^2	1.0E-4
LAMBDA (channel length modulation)	1/V	0
RD (drain resistance)	Ω	0
RS (source resistance)	Ω	0
CGS (zero bias G-S junction capacitance)	F	0
CGD (zero bias G-D junction capacitance)	F	0
PB (gate junction potential)	V	1.0
IS (gate junction saturation current)	A	1E-14
KF (flicker noise coefficient)		0
AF (flicker noise exponent)		1.0
FC (forward-bias depletion capacitance coefficient)		0.5

Examples

```
.MODEL   LGFET   PJF   VTO = -1.5   BETA = 13E-6
+ RD = 50   RS = 50
.MODEL   SMFET   NJF   VTO = 0.9   BETA = 13E-6
+ RD = 200   RS = 200
```

Metal Oxide Semiconductor Field-Effect Transistors

The definition statement for a metal oxide semiconductor field-effect transistor (MOSFET) contains five mandatory data fields and three optional data fields separated by spaces. The format for a MOSFET definition statement is as follows:

```
MXXXXXXX     ND     NG     NS     NB     MNAME
+ L = VAL    W = VAL    AD = VAL    AS = VAL
+ PD = VAL    PS = VAL    NRD = VAL    NRS = VAL
+ OFF    IC = VDS,VGS,VBS
```

The MOSFET definition statement must begin with the key letter M. The M can be followed by up to seven additional letters or numbers to uniquely identify the MOSFET. Next are the ND, NG, NS, and NB fields. These fields describe the node numbers in the circuit to which the drain, gate, source, and the bulk (back gate) of the transistor are connected. "MNAME" is the name of the model describing the MOSFET. The following optional data fields are used to specify device sizing:

Parameter	Units
L (channel length)	m
W (channel width)	m
AD (drain diffusion area)	m^2
AS (source diffusion area)	m^2
PD (drain junction perimeter)	m
PS (source junction perimeter)	m
NRD (drain diffusion squares)	
NRS (source diffusion squares)	

Default values for the data fields listed above are specified on the .OPTIONS card. The optional "OFF," if included, sets the initial terminal voltages of the MOSFET to zero. "IC = VDS, VGS" is used in conjunction with the UIC command on the .TRAN card to set the initial terminal voltages for transient analysis to a voltage other than that determined by the dc bias-point calculation.[40,41]

Metal Oxide Silicon Field-Effect Transistor Model

The model used by SPICE is based on the Shichman-Hodges model.

The following is a list of the MOSFET model parameters and default values:[40,42]

Parameter	Units	Default value
LEVEL (model index)		1
VTO (zero-bias threshold voltage)	V	0
KP (transconductance)	A/V^2	2.0E-5
GAMMA (bulk threshold)	$V^{1/2}$	0
PHI (surface potential)	V	0.6
LAMBDA (MOS1 and MOS2 channel length modulation)	1/V	0
RD (drain resistance)	Ω	0
RS (source resistance)	Ω	0
CBD (zero bias B-D junction capacitance)	F	0
CBS (zero bias B-S junction capacitance)	F	0
PB (gate junction potential)	V	0.8
IS (bulk junction saturation current)	A	1E-14
CGSO (G-S overlap capacitance/meter of channel width)	F/m	0
CGDO (G-D overlap capacitance/meter of channel width)	F/m	0
CGBO (G-B overlap capacitance/meter of channel length)	F/m	0
RSH (S and D diffusion sheet resistance)	Ω/sq	0
CJ (zero-bias bulk junction capacitance)	F/m^2	0
MJ (bulk junction grading coefficient)		0.5
CJSW (zero-bias bulk junction sidewall capacitance per meter of junction perimeter)	F/m	0
MJSW (bulk junction sidewall grading coefficient)	F/m	0
JS (bulk junction saturation current per m^2 of junction area)	A/m^2	0
TOX (oxide thickness)	m	1.0E-7
NSUB (substrate doping)	$1/cm^3$	0
NSS (surface state density)	$1/cm^2$	0
NFS (fast surface state density)	$1/cm^2$	0
TPG (type of gate material) +1 = opposite of substrate −1 = same as substrate 0 = aluminum gate		

XJ (metallurgical junction depth)	m	0
LD (lateral diffusion)	m	0
UO (surface mobility)	cm²/V-S	600
UCRIT (critical field for mobility degradation (MOS2 only)	V/cm	1.0E-4
UEXP (critical field exponent—MOS2 only)		0
UTRA (transverse field coefficient deleted in MOS2)		0
VMAX (maximum carrier drift velocity)	m/s	0
NEFF (total channel charge coefficient MOS2 only)		1.0
XQC (thin-oxide capacitance model flag and coefficient of drain channel charge share (0—0.5))		1.0
KF (flicker noise coefficient)		0
AF (flicker noise exponent)		1.0
FC (forward-bias depletion capacitance coefficient)		0.5
DELTA (width effect on threshold voltage MOS2 and MOS3)		0
THETA (mobility modulation—MOS3)	1/V	0
ETA (static feedback—MOS3)		0
KAPPA (saturation field factor—MOS3)		0.2

Example

```
.MODEL    PFET1    PMOS    LEVEL=2    VTO=-0.4
+ KP=10U    GAMMA=0.53    PHI=0.64    LAMBDA=0.03
```

4.6 Subcircuits

This section will describe, in detail, the subcircuit call and the subcircuit definition. The subcircuit definition describes a block of circuitry that, once described, can be repetitively called just like a component. The subcircuit call is very similar to a passive component definition statement except that instead of a value there is a subcircuit name and there are no optional data fields. This gives SPICE the ability to use a block of circuitry or a behavior model of a block of circuitry just as an integrated circuit would be used in a discrete design. A multiple op amp active filter, for example, would require that the circuit description or model of the op amp be entered only once in a subcircuit definition. The op amp could then be used through a subcircuit call, just like any other component, as many times as necessary.

A subcircuit is a group of cards that defines a block of circuitry, such as an op amp, that can be referred to much like a component. Circuitry defined as a subcircuit is typically used many times in a more global circuit or is a proven circuit of known performance characteristics. The use of subcircuits can significantly reduce the complexity of the SPICE input file. Many subcircuits, such as op amp macromodels, are widely used and can be kept in a library. If many engineers have access to a common subcircuit library, duplication of design effort can be eliminated and the experience on one project can be used on future projects.

Subcircuit Definition

A subcircuit definition begins with a .SUBCKT card and ends with a .ENDS (for END Subcircuit, different from .END card used to end a SPICE file) card. The cards between the .SUBCKT and .ENDS cards define the circuitry that constitutes the subcircuit. The number of cards used to describe a subcircuit is not limited by SPICE and the subcircuit description can contain other subcircuits. The format for the .SUBCKT card is as follows:

```
.SUBCKT     SUBNAME     N1     N2     N3 . . . Nn
```

"SUBNAME" is the unique name of the subcircuit (i.e., OPAMP, RFAMP, FILTER, etc.) and can contain up to eight characters. The node numbers N1 . . . Nn are the nodes (the ground node, 0, cannot be included as one of these nodes) within the subcircuit which are to be connected to the more global circuit. If the subcircuit is considered a component, these are the pin numbers of the component. All node numbers, except 0 (which is always global) and the nodes listed on the .SUBCKT card, subcircuit definitions, and .MODEL cards included within a subcircuit definition are strictly local and are not known outside the subcircuit definition.

The format for a .ENDS card is as follows:

```
.ENDS     SUBNAME
```

"SUBNAME" is optional but should always be included for clarity, especially if the subcircuit definition includes other nested subcircuit definitions.

Subcircuit Calls

Every subcircuit call is unique in the same sense as a component definition is unique. Each time the same subcircuit is called, it is treated as though it were a different component and therefore must have a unique name. The format for a subcircuit call is as follows:

XYYYYYYY N1 N2 N3 . . . Nn SUBNAME

The name of the subcircuit call must start with the key letter X. Seven additional letters or numbers can be used to uniquely define the call. The node numbers N1 . . . Nn are the node numbers of the more global circuit to which the corresponding local subcircuit nodes (from the .SUBCKT card) are to be connected. In other words, the local subcircuit node N1 on the .SUBCKT card is connected to the more global node N1 on the subcircuit call. "SUBNAME" is the name of the subcircuit being called.[43,44]

4.7 Analysis Control

The control cards define the types of analyses, the conditions under which these analyses will be performed, and the timing of the measurements of the circuit response. The analysis control cards are listed in alphabetical order and the use of each is described.[45,46]

.AC

The .AC card invokes an ac analysis. At least one of the independent sources in the circuit must have an ac value. The inclusion of a .AC card is necessary to perform a distortion analysis (.DISTO) or a noise analysis (.NOISE). The format for the .AC card is

.AC TYPE N FSTART FSTOP

There are three types of frequency variation and N is the number of points calculated over the variation.

"TYPE"	"N"
DEC	N points per decade
OCT	N points per octave
LIN	N points between "FSTART" and "FSTOP"

"FSTART" is the starting frequency and "FSTOP" is the ending frequency. The output is specified on a .PLOT or .PRINT card.

Examples

```
.AC   DEC   10    100   10K
.AC   OCT   5     6     600
.AC   LIN   100   1     1K
```

.DC

The .DC card generates dc transfer curves similar to those displayed on a curve tracer. One source can be specified to generate a single curve or two sources can be specified to generate a family of curves. The default dc bias-point calculation is not performed if the dc transfer curves is the only analysis performed. The format for the .DC card is

```
.DC   SRCNAME1   STARTVAL1   STOPVAL1   INC1
+ SRCNAME2   STARTVAL2   STOPVAL2   INC2
```

"SRCNAME1" and optional "SRCNAME2" are the names of independent voltage or current sources in the circuit. "STARTVAL1" and "STARTVAL2" are the current or voltage values from which these sources start their sweep. "STOPVAL1" and "STOPVAL2" are the current or voltage values at which the sweep ends. "INC1" and "INC2" are the increments or steps taken within the sweep range. The output is specified on a .PLOT or .PRINT card.

Examples

```
.DC   VSUPPLY   1.0   10.0   0.5   IBASE   0.1U
+ 1.0U   0.05U
.DC   V1   .01   .1   .005
```

.DISTO

The .DISTO card with a .AC card invokes the small-signal distortion analysis of the circuit. This analysis assumes that one or two signal frequencies, F1 and F2, are driving the input of the circuit. The format of the .DISTO card is

```
.DISTO   RLOAD   INTER   SKW2   REFPWR   SPW2
```

All distortion products are calculated in "RLOAD." A listing of the contributions of the nonlinear devices to total distortion is printed every "INTER" frequency points. If "INTER" is set to zero or omitted, this listing will be omitted. "SKW2" is the ratio of F1 to F2. If "SKW2" is omitted, SPICE assumes F2 = 0.9 F1. "REFPWR" is the reference power for distortion products. The default for "REFPWR" is 1 mW. "SPW2" is the amplitude of F2 and defaults to 1.0 if omitted.

Example

```
.DISTO   R58   2   0.73   1M   0.90
```

.FOUR

The .FOUR card in conjunction with a transient analysis (.TRAN) will perform a discrete Fourier transform (DFT). The dc component and the first nine harmonics will be calculated for the specified output variables. Only one transient analysis and one Fourier analysis is performed per SPICE run. The format of the .FOUR card is

```
.FOUR   FREQ   OV1   OV2 . . . OVn
```

"FREQ" specifies the fundamental frequency and OV1 . . . OVn are the output variables for which the analysis is computed. The Fourier analysis is performed over the interval TSTOP–PERIOD to TSTOP. "TSTOP" is the ending time for the transient analysis and is specified on the .TRAN card. PERIOD = 1/"FREQ." Be sure to set "TMAX" on the .TRAN card to a small enough interval to yield acceptable accuracy in the Fourier analysis.

Examples

```
.FOUR   1MEG   V(17)
.FOUR   10K    V(7)   V(12)   V(28)
```

.IC

The .IC card sets the initial conditions for transient analysis. Unlike the .NODESET card, the voltages specified on the .IC card are used to determine the initial dc conditions, not just to get SPICE headed in the right direction. The .IC card has an impact on the initial dc bias-point calculation whether or not the "UIC" option is selected on the .TRAN card. The format for the .IC card is

.IC V(N1) = VALUE1 . . . V(Nn) = VALUEn

If the UIC option is selected on the .TRAN card, the node voltages on the .IC card are used to determine the semiconductor and capacitor initial conditions. The dc sources that impact the initial conditions of the circuit should also be specified on the .IC card. The initial condition statements on the individual device cards (IC = VALUE) override the values on the .IC card.

If the UIC option is not selected on the .TRAN card, the node voltages specified on the .IC card will be imposed during the initial dc bias-point analysis and then released during transient analysis.

.NODESET

The .NODESET card allows the user to specify node voltages to be applied to the circuit for SPICE's initial pass. The node voltages are then released and SPICE converges to the exact solution. This card is not usually necessary but can be helpful when simulating circuits with more than one stable state to ensure that they come up in the desired state. The format for the .NODESET card is

.NODESET V(N1) = VALUE V(N2) = VALUE . . .

Example

.NODESET V(7) = 6.95 V(22) = 5.0 V(3) = 0

.NOISE

The .NOISE card with an ac analysis (.AC card) invokes the noise analysis. The format of the .NOISE card is

.NOISE OUTV INSRC NUMS

"OUTV" is the output voltage for the noise analysis. "INSRC" is an independent current or voltage source that serves as the input reference. "NUMS" determines the number of frequency points between summary listings. Each summary listing will include output noise, equivalent input noise, and the contributions of each noise generator in the circuit. The equivalent input noise or output noise can be printed (.PRINT) or plotted (.PLOT) in addition to or instead of the summary listing. Set "NUMS" = 0 to omit the summary listing.

Examples

```
.NOISE   V(7)     ISOURCE   10
.NOISE   V(8,6)     V2    0    (no summary listing)
```

.OP

SPICE always performs a dc operating-point analysis (1) if no other analysis is specified, (2) prior to transient analysis to determine the initial conditions, and (3) prior to an ac small-signal analysis to generate linearized models for all nonlinear devices. A dc operating-point analysis is not performed (1) when dc transfer curves are the only analysis in the SPICE run or (2) SPICE is instructed to omit it from the transient analysis because the initial conditions have been supplied.

The .OP card causes an extended dc operating point analysis to be performed. In addition to the default dc bias-point calculation, the extended calculation includes semiconductor parameters including f_t, terminal voltages, terminal currents, and power dissipation. The format for the .OP card is

.OP

.OPTIONS

The .OPTIONS card allows many different program control options to be asserted. The format for the .OPTIONS card is

.OPTIONS OPT1 OPT2 . . . or OPT = X

Some options are specified by simply listing their name and others by setting the option equal to a positive value (X). The options can be input in any order.

Option name	Description
ABSTOL = X	Sets absolute current error (default is 1.0E-12)
ACCT	Lists accounting and run time statistics
CHGTOL	Sets absolute charge error (default is 1.0E-14)
CPTIME = X	Sets maximum CPU time for run
DEFAD = X	Sets MOS drain diffusion area (default = 0.0)
DEFAS = X	Sets MOS source diffusion area (default = 0.0)
DEFL = X	Sets MOS channel length (default = 1.0E-4 m)
DEFW = X	Sets MOS channel width (default = 1.0E-4 m)
GMIN = X	Sets minimum conductance (default = 1.0E-12)
ITL1 = X	Sets dc iteration limit (default = 100)
ITL2 = X	Sets dc transfer curve iteration limit (default = 50)
ITL3 = X	Sets lower transient analysis iteration limit (default = 4)
ITL4 = X	Sets transient analysis time-point iteration limit (default = 4)
ITL5 = X	Sets total transient analysis iteration limit (default = 5000); set XTL5 = 0 to omit this test
LIMPTS = X)	Sets maximum print/plot points (default = 201).
LIST	Prints listing of input data
LIMTIM = X	Sets reserve CPU time for plots (default = 2 sec)
LVLTIM = X	Time-step control X = 1—Iteration time-step control X = 2—Truncation-error time-step (default = 2) If METHOD = GEAR and MAXORD $>$ 2 SPICE sets X = 2
MAXORD = X	Sets maximum integration order if METHOD = GEAR. $2 \leq X \geq 6$ (default = 2)
METHOD = X	Sets numerical integration method X = GEAR or X = TRAPEZOIDAL) (default = TRAPEZOIDAL)
NODE	Node table printed
NOMOD	Model parameters are not printed
NOPAGE	No page ejects
NUMDGT = X	Sets significant digits for output $(1 \leq X \geq 7)$ (default = 4); NUMDGT is independent of the error tolerance

OPTS	Option values are printed
PIVREL = X	Sets ratio between the largest column entry and acceptable pivot value (default = 1.0E-3)
PIVTOL = X	Sets absolute minimum value for a matrix entry to be acceptable as a pivot (default = 1.0E-13)
RELTOL = X	Sets the relative error tolerance of SPICE (default = 0.001 (0.1%))
TNOM = X	Sets nominal temperature (default = 27°C (300°K))
TRTOL = X	Sets transient error tolerance (default = 7.0)
VNTOL = X	Sets absolute voltage error (default = 1.0E-6)

.SENS

The .SENS card invokes a sensitivity analysis of each specified node voltage or current with respect to the value of every component in the circuit. Needless to say, this analysis can generate a very large volume of output for large circuits. Discretion is important when specifying outputs. The format of the .SENS card is

```
.SENS    OV1    OV2 . . . OVn
```

Example

```
.SENS    V(5,3)    V(2)    I(VSUPPLY)
```

.TEMP

The .TEMP card is an optional card that allows the specification of any number of temperatures greater than −223.0°C at which to run the simulation. If this card is omitted, the simulation will be run at T = TNOM as specified on the .OPTIONS card. All model parameters are assumed to be given for T = TNOM. Temperature variations are calculated from that reference point. The format of the .TEMP card is as follows:

```
.TEMP    T1    T2 . . . Tn
```

Example

```
.TEMP    -25.0    0.0    25.0    50.0    75.0
```

.TF

The .TF card invokes the small-signal dc transfer function. The format for the .TF card is

.TF OUTVAR INSRC

"OUTVAR" is the output node voltage or current and "INSRC" is the input current or voltage source. SPICE calculates the dc small-signal transfer function, input impedance at the source, and the output impedance.

Examples

```
.TF   V(3)      IIN
.TF   V(7,10)   VSENSE
.TF   I(VIN)    V3
```

.TRAN

The .TRAN card invokes a transient analysis. Only one transient analysis can be performed in a given SPICE run. A transient analysis is necessary for SPICE to perform a Fourier (.FOUR) analysis. The format for a .TRAN card is

.TRAN TSTEP TSTOP TSTART TMAX UIC

"TSTEP" is the increment at which data points are taken to either be printed (.PRINT) or plotted (.PLOT). "TSTOP" is the ending time of the transient analysis. The actual transient analysis begins at time zero, but data points are not stored until TSTART. If TSTART is omitted, SPICE begins storing data at time-zero. "TMAX" specifies the maximum step size and allows the user to guarantee the step size smaller than TSTEP. The default computational step size is the smaller of (TSTOP − TSTART)/50 or TSTEP.

The optional key word UIC (use initial conditions) tells SPICE to omit the initial dc bias-point analysis and use the initial conditions specified on either the semiconductor and capacitor device cards (these values have precedence over the values specified on the .IC card) or those specified on the .IC card.

Examples

```
.TRAN    100N    1U    500N    50N
.TRAN    50U     1M    750U    UIC
```

4.8 Output Control

There are four cards that control the output produced by the standard versions of SPICE. The .OPTIONS card has some output control capability which was discussed in Section 4.7. The other three output control cards are the .PLOT, .PRINT, and .WIDTH cards. The many commercially available graphics postprocessing packages add a significantly more robust set of analysis tools than those available with standard SPICE and should be investigated by the serious SPICE user. The discussion here will be limited to the standard tool set available with Berkeley SPICE.

.PLOT

The .PLOT card allows the analysis results of up to eight variables to be plotted with standard line printer characters. The format of the .PLOT card is as follows:[47,48]

```
.PLOT    PLTYPE    OV1    PLO1,PHI1 . . . OV8    PLO8,PHI8
```

".PLOT" tells SPICE to plot a graph of the specified variables. "PLTYPE" specifies the type of analysis that generated the data to be plotted. PLTYPE can be one of the following: AC, DC, DISTO, NOISE, or TRAN. OV1 . . . OV8 are the voltage or current output variables to be plotted. PLO and PHI are the optional plot limits. All output variables to the left of these plot limits will be plotted using these limits. If the plot limits are omitted, the plot will be scaled to display all of the data. SPICE will generate multiple scales, in this case, if there are large differences between two or more output variables. If multiple output variables are plotted, the value of the first one specified will be printed at each point and overlaps on the plot will be indicated with a letter X.

The output variables have one format for data generated by AC, DC, and TRAN, and a different format for DISTO and NOISE.

Output Variable Format for AC, DC, and TRAN

A voltage specified as an output variable for DC and TRAN has the following format:

V(N1, N2).

A current specified as an output variable for DC and TRAN has the following format:

I(VXXXXXXX)

The "V" or "I" indicates that the output variable is a dc voltage or current. To show the results of an ac analysis, the "V" or "I" is replaced by one of the following:

Voltage	Current	Output
VR	IR	Real part of a complex value
VI	II	Imaginary part of a complex value
VM	IM	Magnitude of a complex value
VP	IP	Phase of a complex value
VDB	IDB	$20 \log_{10}$ (VM) or $20 \log_{10}$ (IM)

After the type of voltage output has been specified, the node numbers between which the measurement is to be made are specified. N2 is the reference node. The output variable, N1, is measured with respect to N2. If N2 and the comma are omitted, ground (node 0) is assumed by SPICE to be the reference node.

After the type of current output has been specified, the voltage source through which this current is to be measured is specified by listing its name in parentheses. Positive current flows into the source from the positive node.

Output Variable Format for NOISE and DISTO

An measurement specified as an output variable for NOISE and DISTO has the following format:

OV(X)

OV is one of the following:

OV	Description
ONOISE	Output noise
INOISE	Equivalent input noise
HD2	Magnitude of 2F1 (F2) not present
HD3	Magnitude of 3F1 (F2) not present
SIM2	Magnitude of F1 + F2
DIM2	Magnitude of F1 − F2
DIM3	Magnitude of 2F1 − F2

The "X" is replaced by one of the following:

R	Real part of a complex value
I	Imaginary part of a complex value
M	Magnitude of a complex value
P	Phase of a complex value
DB	20 $\log_{10}$ (VM)

If an output variable, OV, is specified and the (X) is omitted, the default is magnitude (M).

There are no limitations on the number of .PLOT cards for a given analysis.

Examples

```
.PLOT    TRAN     V(3)      V(4,7)
.PLOT    AC     VM(7)     VP(23)
.PLOT    AC     VDB(8)     VP(8)
.PLOT    NOISE     ONOISE(DB)     INOISE(DB)
```

.PRINT

This card causes SPICE to print a listing of every value of up to eight specified output variables. The format of the .PRINT card is as follows:[49,50]

```
.PRINT     PRTYPE     OV1 . . . OV8
```

The syntax of the .PRINT card is the same as that for the .PLOT card. "PRTYPE" corresponds to "PLTYPE" and the plot limits, PLO and PHI, are omitted since the values are being printed in a list instead of being plotted. There are no limitations on the number of .PRINT cards for a given analysis.

Examples

```
.PRINT   TRAN     V(3)     V(4,7)
.PRINT   AC     VM(7)     VP(23)
.PRINT   AC     VDB(8)     VP(8)
.PRINT   NOISE     ONOISE(DB)     INOISE(DB)
```

.WIDTH

The .WIDTH card specifies the input and output column widths.

```
.WIDTH     IN = COLNUM     OUT = COLNUM
```

The IN parameter is the last column (COLNUM = column number) read from each line of input to SPICE and the OUT parameter specifies the print width of the output and can be either 80 or 133 columns.[50,51]

Example

```
.WIDTH OUT=80
```

4.9 Typical Model Parameters

This section presents some "typical" semiconductor device model parameters found on semicustom arrays (see Tables 4-1 through 4-5). These models will be used extensively in Chapter 5 to illustrate the performance of example circuits. It is important to note that these model parameters, while representative of semicustom arrays in general, do not necessarily correspond to the parameters provided by any manufacturer. These models can be used for rough design work and feasibility studies. However, up-to-date vendor-supplied models should be used for detailed design work and for verification of circuit performance prior to integration. Remember that simulation accuracy depends on the accuracy of the models used to perform the simulation.

Table 4-1
Small-Geometry *npn* Transistor Model (SGNPN)[a]

Parameter	Model value	Units	Default value
IS	5.0E-16	A	1.0E-16
BF	150		100
NF			1.0
VAF	200	V	Infinite
IKF	1.0E-3	A	Infinite
ISE	1.0E-13	A	0
NE			1.5
BR			1.0
NR			1.0
VAR		V	Infinite
IKR		A	Infinite
ISC	3.0E-12	A	0
NC			2.0
RB	150	Ω	0
IRB		A	Infinite
RBM		Ω	RB
RE	2.0	Ω	0
RC	50	Ω	0
CJE	2.5E-12	F	0
VJE	0.6	V	0.75
MJE			0.33
TF	0.33E-9	sec	0
XTF			0
VTF		V	Infinite
ITF		A	0
PTF		deg	0
CJC	2.0E-12	F	0
VJC		V	0.75
MJC			0.33
XCJC			1.0
TR	10.0E-9	sec	0
CJS	2.0E-12	F	0
VJS		V	0.75
MJS			0
XTB	2.0		0
EG		eV	1.11
XTI			3.0
KF			0
AF			1.0
FC			0.5

[a]**The .MODEL card for SGNPN is as follows:**

```
.MODEL SGNPN NPN(IS=5.0E-16 BF=150 VAF=200 IKF=1.0E-3
+ ISE=1E-13 ISC=3.0E-12 RB=150 RE=2.0 RC=50 CJE=2.5E-12
+ VJE=0.6 TF=0.33E-9 CJC=2.0E-12 TR=10.0E-9 CJS=2.0E-12
+ XTB=2.0)
```

Table 4-2
Large-Geometry *npn* Transistor Model (LGNPN)[a]

Parameter	Model value	Units	Default value
IS	5.0E-16	A	1.0E-16
BF	150		100
NF			1.0
VAF	150	V	Infinite
IKF	10.0E-3	A	Infinite
ISE	1.0E-13	A	0
NE			1.5
BR			1.0
NR			1.0
VAR		V	Infinite
IKR		A	Infinite
ISC	3.0E-12	A	0
NC			2.0
RB	2.0	Ω	0
IRB		A	Infinite
RBM		Ω	RB
RE	1.0	Ω	0
RC	10.0	Ω	0
CJE	12.0E-12	F	0
VJE	0.6	V	0.75
MJE			0.33
TF	0.5E-9	sec	0
XTF			0
VTF		V	Infinite
ITF		A	0
PTF		deg	0
CJC	10.0E-12	F	0
VJC		V	0.75
MJC			0.33
XCJC			1.0
TR	15.0E-9	sec	0
CJS	10.0E-12	F	0
VJS		V	0.75
MJS			0
XTB	2.0		0
EG		eV	1.11
XTI			3.0
KF			0
AF			1.0
FC			0.5

[a]The .MODEL card for LGNPN is as follows:

```
.MODEL LGNPN NPN(IS=5.0E-16 BF=150 VAF=150 IKF=10.0E-3
+ ISE=1.0E-13 ISC=3.0E-12 RB=2.0 RE=1.0 RC=10 CJE=12.0E-12
+ VJE=0.6 TF=0.5E-9 CJC=10.0E-12 TR=15.0E-9 CJS=10.0E-12
+ XTB=2.0)
```

Table 4-3
Lateral *pnp* Transistor Model (LATPNP)[a]

Parameter	Model value	Units	Default value
IS		A	1.0E-16
BF	70		100
NF			1.0
VAF	80	V	Infinite
IKF	50.0E-6	A	Infinite
ISE	1.0E-13	A	0
NE			1.5
BR			1.0
NR			1.0
VAR		V	Infinite
IKR		A	Infinite
ISC	3.0E-12	A	0
NC			2.0
RB	300	Ω	0
IRB		A	Infinite
RBM		Ω	RB
RE	10.0	Ω	0
RC	200	Ω	0
CJE	1.0E-12	F	0
VJE	0.65	V	0.75
MJE			0.33
TF	33.0E-9	sec	0
XTF			0
VTF		V	Infinite
ITF		A	0
PTF		deg	0
CJC	2.0E-12	F	0
VJC		V	0.75
MJC			0.33
XCJC			1.0
TR	75.0E-9	sec	0
CJS	2.0E-12	F	0
VJS		V	0.75
MJS			0
XTB	2.0		0
EG		eV	1.11
XTI			3.0
KF			0
AF			1.0
FC			0.5

[a] **The .MODEL card for LATPNP is as follows:**

```
.MODEL LATPNP PNP(BF=70 VAF=80 IKF=50.0E-6 ISE=1E-13
+ ISC=3.0E-12 RB=300 RE=10.0 RC=200 CJE=1.0E-12
+ VJE=0.65 TF=33.0E-9 CJC=2.0E-12 TR=75.0E-9 CJS=2.0E-12
+ XTB=2.0)
```

Table 4-4
Small-Geometry Vertical *pnp* Transistor Model (SVERTPNP)[a]

Parameter	Model value	Units	Default value
IS	1.0E-15	A	1.0E-16
BF	90		100
NF			1.0
VAF	80	V	Infinite
IKF	1.0E-3	A	Infinite
ISE	1.0E-13	A	0
NE			1.5
BR			1.0
NR			1.0
VAR		V	Infinite
IKR		A	Infinite
ISC	3.0E-12	A	0
NC			2.0
RB	300	Ω	0
IRB		A	Infinite
RBM		Ω	RB
RE	10.0	Ω	0
RC	100	Ω	0
CJE	1.0E-12	F	0
VJE	0.65	V	0.75
MJE			0.33
TF	20.0E-9	sec	0
XTF			0
VTF		V	Infinite
ITF		A	0
PTF		deg	0
CJC		F	0
VJC		V	0.75
MJC			0.33
XCJC			1.0
TR	60.0E-9	sec	0
CJS	2.0E-12	F	0
VJS		V	0.75
MJS			0
XTB	2.0		0
EG		eV	1.11
XTI			3.0
KF			0
AF			1.0
FC			0.5

[a] **The .MODEL card for SVERTPNP is as follows:**

```
.MODEL SVERTPNP PNP(IS=1.0E-15 BF=90 VAF=80 IKF=1.0E-3
+ ISE=1.0E-13 ISC=3.0E-12 RB=300 RE=10.0 RC=100
+ CJE=1.0E-12 VJE=0.65 TF=20.0E-9 TR=60.0E-9 CJS=2.0E-12
+ XTB=2.0)
```

Table 4-5
Large-Geometry Vertical *pnp* Transistor Model (LVERTPNP)[a]

Parameter	Model value	Units	Default value
IS	1.0E-14	A	1.0E-16
BF	90		100
NF			1.0
VAF	80	V	Infinite
IKF	5.0E-3	A	Infinite
ISE	1.0E-13	A	0
NE			1.5
BR			1.0
NR			1.0
VAR		V	Infinite
IKR		A	Infinite
ISC	3.0E-12	A	0
NC			2.0
RB	100	Ω	0
IRB		A	Infinite
RBM		Ω	RB
RE	2.0	Ω	0
RC	100	Ω	0
CJE	5.0E-12	F	0
VJE	0.65	V	0.75
MJE			0.33
TF	30.0E-9	sec	0
XTF			0
VTF		V	Infinite
ITF		A	0
PTF		deg	0
CJC		F	0
VJC		V	0.75
MJC			0.33
XCJC			1.0
TR	75.0E-9	sec	0
CJS		F	0
VJS		V	0.75
MJS			0
XTB	2.0		0
EG		eV	1.11
XTI			3.0
KF			0
AF			1.0
FC			0.5

[a] **The .MODEL card for LVERTPNP is as follows:**

```
.MODEL LVERTPNP PNP(IS=1.0E-14 BF=90 VAF=80 IKF=5.0E-3
+ ISE=1E-13 ISC=3.0E-12 RB=100 RE=2.0 RC=100 CJE=5.0E-12
+ VJE=0.65 TF=30.0E-9 TR=75.0E-9 XTB=2.0)
```

References

1. A. Vladimirescu, K. Zhang, A. R. Newton, D. O. Pederson, A. Sangiovanni-Vincentelli, *SPICE 2G User's Guide*, University of California Berkeley, California, 1981, Section 5.3.
2. Vladimirescu *et al.*, (1981), Section 3.0.
3. L. Meares, C. Hymowitz, *Simulating With SPICE*, intusoft, San Pedro, CA, 1988, pp. 2–15.
4. A. Vladimirescu *et al.* (1981), Section 4.0.
5. A. Vladimirescu *et al.* (1981), Section 9.15.
6. A. Vladimirescu *et al.* (1981), Section 9.16.
7. A. Vladimirescu *et al.* (1981), Section 6.1.
8. L. Meares, *et al.* (1988), pp.1-10.
9. A. Vladimirescu *et al.* (1981), Section 6.2.
10. L. Meares, *et al.* (1988), pp.1-11.
11. L. Meares, C. Hymowitz, (1988), pp.1-12.
12. A. Vladimirescu *et al.* (1981), Section 6.3.
13. L. Meares, C. Hymowitz, (1988), pp.1-13.
14. A. Vladimirescu *et al.* (1981), Section 6.4.
15. L. Meares, C. Hymowitz, (1988), pp.1-14.
16. A. Vladimirescu *et al.* (1981), Section 6.5.
17. A. Vladimirescu *et al.* (1981), Section 6.6.
18. L. Meares, C. Hymowitz, (1988), pp.1-22.
19. A. Vladimirescu *et al.* (1981), Section 6.7.
20. L. Meares, C. Hymowitz, (1988), pp.1-19.
21. A. Vladimirescu *et al.* (1981), Section 6.8.
22. L. Meares, C. Hymowitz, (1988), pp.1-20.
23. A. Vladimirescu *et al.* (1981), Section 6.9.
24. L. Meares, C. Hymowitz, (1988), pp.1-21.
25. A. Vladimirescu *et al.* (1981), Appendix B.
26. L. Meares, C. Hymowitz, (1988), pp.1-15 through 1-18.
27. A. Vladimirescu *et al.* (1981), Section 6.10.
28. A. Vladimirescu *et al.* (1981), Section 7.5.
29. L. Meares, C. Hymowitz, (1988), pp.1-41.
30. A. Vladimirescu *et al.* (1981), Section 7.1.
31. L. Meares, C. Hymowitz, (1988), pp.1-23 through 1-25.
32. A. Vladimirescu *et al.* (1981), Section 7.6.
33. L. Meares, C. Hymowitz, (1988), pp.1-26 through 1-32.
34. A. Vladimirescu *et al.* (1981), Section 7.2.
35. A. Vladimirescu *et al.* (1981), Section 7.7.
36. A. Vladimirescu *et al.* (1981), Appendix C.
37. L. Meares, C. Hymowitz, (1988), pp.1-33 through 1-34.
38. A. Vladimirescu *et al.* (1981), Section 7.3.
39. A. Vladimirescu *et al.* (1981), Section 7.8.
40. L. Meares, C. Hymowitz, (1988), pp.1-35 through 1-39.
41. A. Vladimirescu *et al.* (1981), Section 7.4.
42. A. Vladimirescu *et al.* (1981), Section 7.9.
43. L. Meares, C. Hymowitz, (1988), pp.1-40.
44. A. Vladimirescu *et al.* (1981), Section 8.0.

45. A. Vladimirescu *et al.* (1981), Section 9.0.
46. L. Meares, C. Hymowitz, (1988), pp.1-42 through 1-53.
47. A. Vladimirescu *et al.* (1981), Section 9.16.
48. L. Meares, C. Hymowitz, (1988), pp.1-49.
49. A. Vladimirescu *et al.* (1981), Section 9.15.
50. L. Meares, C. Hymowitz, (1988), pp.1-48.
51. A. Vladimirescu *et al.* (1981), Section 9.2.

CHAPTER 5

Circuit Design

5.0 Introduction to Linear Integrated Design

The purpose of this chapter is to provide an electrical engineer, experienced in discrete component-level circuit design, the necessary concepts to design in the integrated environment. This presentation is meant to illustrate both design philosophy and technique. The chapter starts with building block circuits and later uses them to build more complex circuitry. The use of SPICE simulation for verifying circuit performance is presented in Sections 5.3 and 5.5. All circuitry described in this chapter is intended for illustrative purposes only. No guarantees are made or implied regarding the suitability of any circuits or device models for a particular application, semicustom product, or vendor. The intent is to give a design engineer knowledge of and exposure to the process of IC design. This presentation is not meant to be complete in the sense of covering the topics in infinite detail, but rather it is a general exposure with enough detail to make the material meaningful and useful. Reference is made to other publications where more detail is available should that be desirable. These references should save research time for the detail-oriented reader and spare the more casual reader the frustration of plowing through unnecessary detail.

Integrated design is different from discrete design because of an inversion in philosophy. Discrete transistor level design approaches are strongly influenced by the following facts:

- Passive components are less expensive than active components.
- Components that match or ratio closely are difficult to realize.
- Low absolute value tolerance components are readily available.
- Different technologies can be easily intermixed.
- Component values are easily changed to fine-tune the design.
- Inductors and large-value capacitors are available.
- Large currents and power dissipation can be dealt with easily.

Integrated design approaches are strongly influenced by the following facts:

- Active components are less expensive than passive components.
- Components that match or ratio are readily available.
- Components have wide absolute value tolerances.
- Manufacturing technologies are not easily intermixed.
- Design changes are expensive and time-consuming.
- Inductors and large-value capacitors are not available.
- Large currents and high power dissipation are hard to deal with.

The result is that discrete transistor-level designs tend to emphasize reliance on resistive biasing, tight absolute value tolerances, multiple technologies, inductors, and large-value capacitors where appropriate. The relatively high currents and power dissipation due to the resistive biasing is not a problem since it is handled by a large number of components spread over a relatively large area. Component value changes, either by substitution or the "tweaking" of adjustable components, can be readily achieved to fine-tune a design.

Integrated design, on the other hand, focuses on current-source biasing, component ratios, and matching. Multiple technologies, inductors, and large-value capacitors are impractical on a single chip. Power dissipation is a problem and is kept to a minimum. Component substitution and "tweaking" are impractical in an integrated design. Every possible effort must be made to ensure first-time design success. Integrated design makes heavy use of circuit simulation with computers and breadboards. Circuit simulation will be discussed at length and examples given as circuit design techniques are explored.

Integrated design has both advantages and limitations when compared to discrete design. To maximize the strengths of integrated design for a particular application, it is important to realize the circuit in a way that maximizes the strengths of integrated design and downplays its limitations. The following guidelines will help a designer optimize the design approach:

• **Carefully define circuit performance requirements of the prospective IC with respect to its interfaces.** Think of the "function" of the IC in integrated terms. Write a formal specification considering signal levels, impedances, supply voltages, drive requirements, and fault conditions. Develop test requirements as the circuit is being defined, not as an afterthought.

• **Design-in testability.** ICs that are easy to test are less expensive to manufacture, have a higher incoming quality level, and are more reliable since marginalities are easier to detect.

• **Use a modularized approach to the design.** This is especially true on more complex circuits. Common circuit blocks minimize design and simulation efforts and improve the chances of first-time success. A relatively small repetitively used block of circuitry will be better understood and more carefully designed than a large rambling circuit. This also allows the use of existing vendor-supplied macro circuits. These circuits are usually "tried and true" and have a great deal of characterization data. Many vendors can supply macro circuits in packaged form for use in breadboarding or for further characterization. Don't reinvent the wheel.

• **Minimize the use of passive components.** One of the major strengths of integrated design is the availability of large numbers of active devices.

• **Avoid dependence on absolute values.** The absolute values of integrated resistors found on semicustom arrays vary by a significant amount. Matching, ratioing, and thermal tracking are inherent strengths of integrated design. Reliance on absolute values can be disastrous. Use an external resistor if an absolute value dependency is necessary. This will cost an external component, 1 to 2 pins on the package, and some PC board real estate but will be significantly less expensive than a custom IC with a poor production yield.

• **Avoid the use of large-value reactive components.** These devices cannot be integrated and pose potential hazards to an IC when used externally. Large capacitors, for example, can, under the right conditions, destroy an IC when its power supply is suddenly removed. Special circuit design techniques can be used to minimize these possibilities.

• **Make a provision in the design to allow for the external "tweaking" of critical circuit functions.** The external control of parameters such as bias currents, frequency response, voltage references, and thresholds allows the design to be fine-tuned after integration and adapted to changing system-level requirements without a redesign of the IC.

• **Consider the economics of partitioning a very complex circuit function into two relatively simple custom ICs as opposed to one large complex custom IC.** The combined cost of two relatively small ICs may be less than that of one large one due to higher production yields and lower package costs. Testability could be significantly enhanced. Second sources may be easier to find. The reduction of PC board layout complexity could significantly reduce the design cycle.

• **Review the design with the custom IC vendor.** The vendor's engineering staff has a great deal of experience in custom IC design and production. They can identify potential problems they have experienced in the past and save you a great deal of trouble.

5.1 Current Sources and Biasing Techniques

Next to the transistor, current sources are the most fundamental concept in modern integrated design. Most circuitry on a linear IC functions by modulating, steering, or switching currents. Bipolar transistors are, fundamentally, current-operated devices (as opposed to FETs, which are voltage operated). It is therefore reasonable to expect that, when designing with bipolar transistors, you will be designing with currents. Depending on their desired usage, currents can be fixed, controlled by signals, or absolute temperature. The latter case has many uses including precision voltage references. Current sources, due to their high output impedance, substitute as high-value resistors for loads on gain stages and make possible gains in the range of a million. To realize current sources, very close matching, ratioing, and thermal tracking of transistors are necessary.

The term "current source" is often used to refer to both a circuit from which positive current flows (a source) and a circuit into which positive current flows (a sink). These circuits are also called "turnarounds" because current pushed into (or pulled from) one side is pulled from (or pushed from) the other side.

Current sources can be a very simple two-transistor circuit or a relatively complex arrangement of resistors and cascoded transistors. The "simple" circuits perform adequately for routine applications, while the more complex circuits compensate for the mismatches and offsets caused by finite beta and output impedance imbalances. One of the circuit designer's tasks is to choose a particular circuit configuration for an application that will offer adequate performance without adding unnecessary complexity.

As described in Chapter 3, there is a correlation between base-emitter voltage and collector current on a given transistor or between

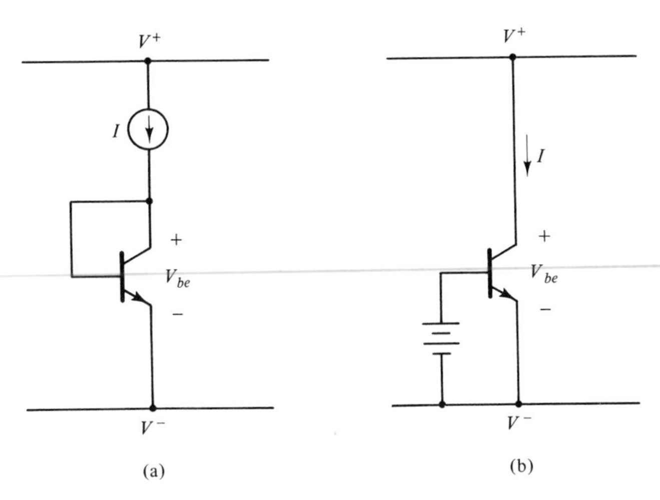

Figure 5-1 $V_{be}V_sV_c$ relationship. (*a*) Forcing current causes V_{be}; (*b*) forcing V_{be} causes current.

closely matched or ratioed transistors at a given temperature. Neglecting error terms, Figure 5-1(*a*) shows that, for a diode-connected transistor, forcing a current through the device causes a certain V_{be}. Conversely, applying that V_{be} to the same (or closely matched) transistor causes the same current to flow. If these two closely matched devices are connected, as shown in Figure 5-2, the current flowing in the diode is reflected in the collector current of the transistor. This works equally as well for *pnp* transistors, as illustrated in Figure 5-3 with the following caveat: *pnp* transistors have lower current capacity, lower beta, and lower output impedance than *npn* transistors. Be aware that significant

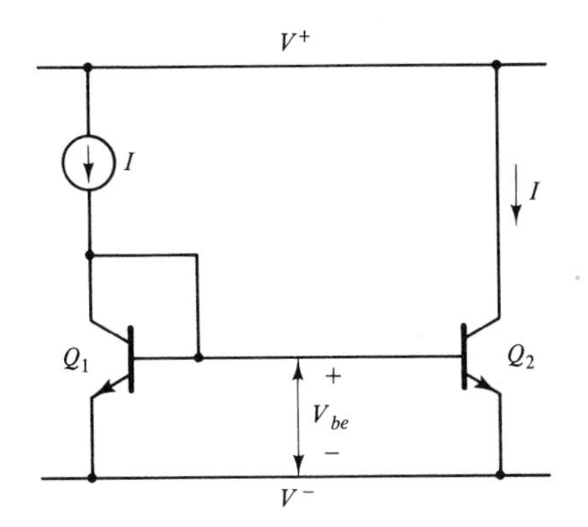

Figure 5-2 *npn* current mirror.

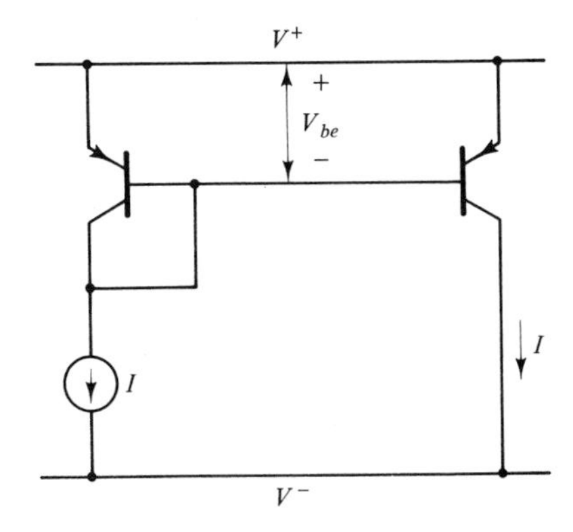

Figure 5-3 *pnp* current mirror.

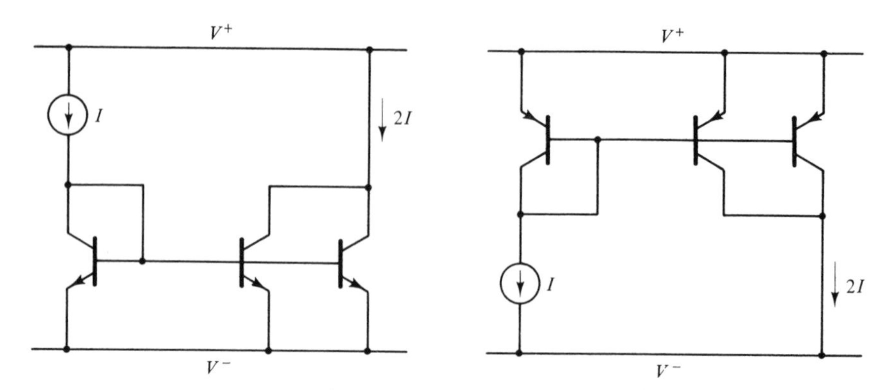

Figure 5-4 Parallel output transistors increase output current.

errors can be associated with *pnp* current sources. More about this later.

For ideal transistors (infinite output impedance and beta) with equal emitter area for *npns* and equal emitter periphery for *pnps*, the output current will equal the input current. If two identical geometry transistors are paralleled on the output side of the current mirror, the output current will be double that of the input side, as shown in Figure 5-4. If two parallel diode-connected transistors are on the input side, as in Figure 5-5, the output current will be half that of the reference side. This scheme works for ratios of 3, 4, 5, and so on (assuming ideal transistors). In full custom designs, the geometry of the transistors can be

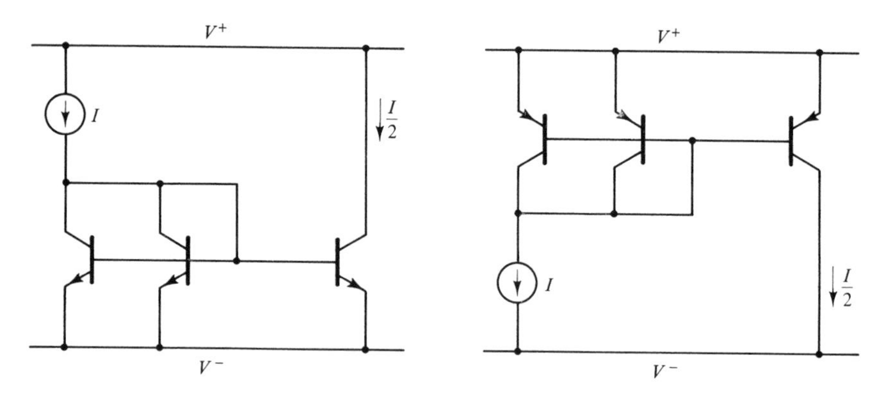

Figure 5-5 Parallel diodes decrease output current.

scaled to yield ratios such as 3.75 : 1. In semicustom design, where transistor geometries are fixed, either integer ratios must be used or emitter ballast resistors must be added to fine-tune the ratio.

Emitter ballast resistors serve two functions in current sources. First, they allow current ratios to be adjusted in other than integer increments. Second, they provide negative feedback to prevent "base current hogging" by one transistor on a bias string. Figure 5-6(*a*) demonstrates how output current can be increased and Figure 5-6(*b*) dem-

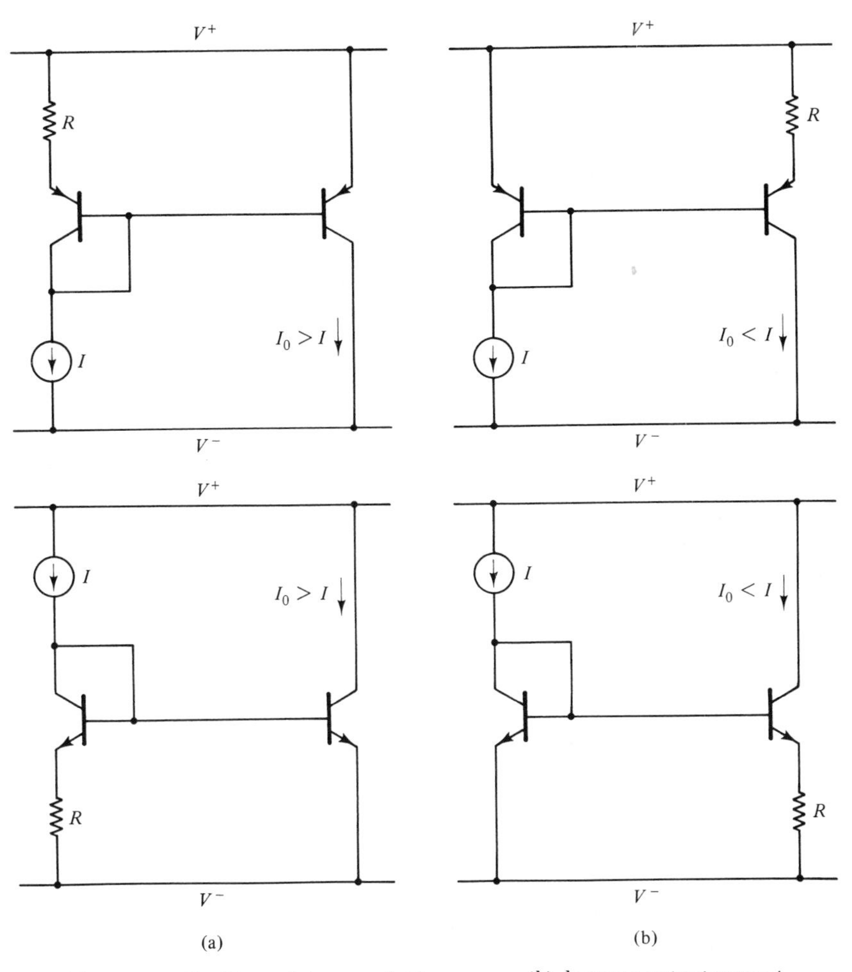

Figure 5-6 Emitter resistors can (*a*) increase or (*b*) decrease output current.

onstrates how output current can be decreased with respect to the reference current using emitter ballast resistors.

Next, the relationship used to calculate the emitter ballast resistor value will be developed. Recall that the current through a diode can be described as (neglecting the -1)

$$I = I_s \exp\left(\frac{qV_{be}}{kT}\right)$$

The ratio of two currents can be expressed as

$$\frac{I_1}{I_2} = \frac{I_s \exp\left(\frac{qV_{be1}}{kT}\right)}{I_s \exp\left(\frac{qV_{be2}}{kT}\right)}$$

Canceling out I_s, taking the natural log of both sides and rearranging yields

$$\Delta V_{be} = V_{be1} - V_{be2} = \frac{kT}{q} \ln\left(\frac{I_1}{I_2}\right)$$

This equation expresses the necessary difference in V_{be} required for two identical transistors at the same temperature to produce a current mirror with the ratio I_1/I_2.

Example: Calculate the necessary value of R in Figure 5-7 to make $I_2 = 40\ \mu A$. Assume infinite beta and infinite output impedance for the identical transistors Q_1 and Q_2.

$$\Delta V_{be} = \frac{kT}{q} \ln\left(\frac{I_1}{I_2}\right) = 0.0259 \ln\left(\frac{100\ \mu A}{40\ \mu A}\right)$$

$$\Delta V_{be} = 23.73\ mV$$

$$R = \frac{23.73\ mV}{40\ \mu A} = 593.3\ \Omega$$

Note: If this same value of resistor were placed in the emitter of Q_1 and I_1 was 40 μA, I_2 would be 100 μA.

"Current hogging" occurs when several transistors are biased by one diode (or voltage reference) and one or more of the current source transistors are mismatched due to small differences in transistor ge-

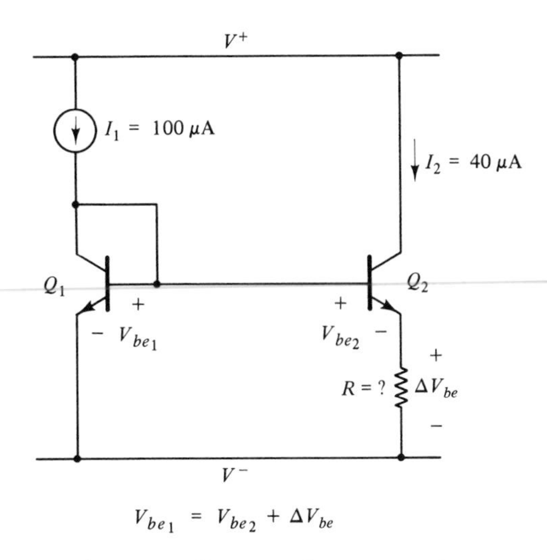

Figure 5-7 Resistor value computation.

ometry or temperature. Temperature differences can readily occur due to voltage differences on the collectors (local heating of a transistor due to its own power dissipation) or by one transistor's proximity to another power dissipating device. The result is that the hotter transistor has a lower V_{be} than the others on the common bias string for a given current. The hotter transistor will take (hog) a disproportionate base current from the bias string, thus disrupting the ratios of the currents. A cure for this undesirable action is to use emitter ballast resistors in the emitters of all transistors on the bias string. A rule of thumb is to drop approximately 100 mV across the emitter resistor. The emitter resistor provides negative feedback that counteracts the current hogging action of a warmer transistor. As the V_{be} becomes smaller due to heating, the increased current through the transistor develops a larger voltage across the emitter resistor, thus tending to turn off the transistor. The use of these ballast resistors is most important on long bias lines where transistors may be physically far apart on a die or where one or more of the transistors are relatively closer to a power-dissipating device such as an output transistor. A negative impact of these resistors is that they reduce the voltage compliance of the current source by the amount of voltage drop on the resistor. On the other hand, emitter ballast resistors improve the output impedance of the current source.

Figure 5-8 illustrates a typical current-source bias arrangement that might be found in an IC. Notice how one reference current is used to provide bias for the whole circuit. The bias determining resistor could be located on the chip or could be an external resistor, trans-

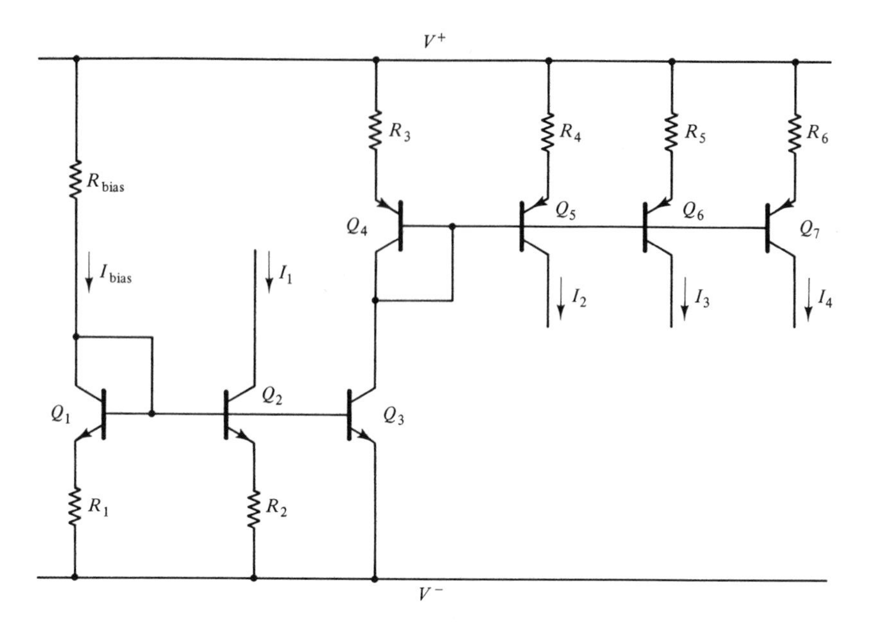

Figure 5-8 Typical current-source bias circuit.

ducer, or current source. So called "programmable op amps" have the bias resistor external to the chip to allow the user to tailor the supply current, speed, frequency response, and output drive for a particular application. The device can be put into an idle mode by reducing the bias current to a minimum when the circuit is not needed, thus saving power-supply current. This feature can be attractive for battery-powered applications.

An interesting variation of the basic current source is the peaking current source illustrated in Figure 5-9. I_1 flowing through R creates a ΔV_{be} between Q_1 and Q_2. As I_1 is increased from zero, I_0 approximately tracks I_1 until the voltage across R (ΔV_{be}) approaches $kT/q = 25.9$ mV. At this point the V_{be} of Q_2 starts to decrease significantly with respect to that of Q_1 and Q_2 begins to turn off, causing I_0 to decrease. The result is shown in Figure 5-10. The transfer function between I_1 and I_0 peaks when $I_1 = kT/q$ and therefore when $I_0 = I_1 \exp(-1)$. This can be shown mathematically as follows.

Assuming:

Q_1 and Q_2 are equal geometries and are well matched
Beta errors are neglected
r_0 errors are neglected

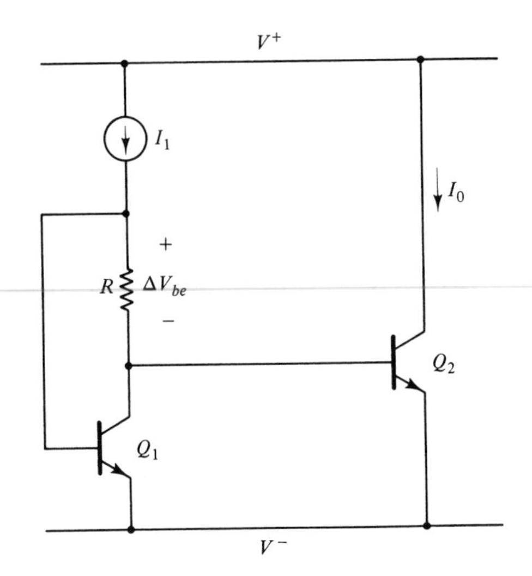

Figure 5-9 Peaking current source.

The difference in V_{be} between Q_1 and Q_2 is

$$\Delta V_{be} = \frac{kT}{q} \ln\left(\frac{I_1}{I_0}\right)$$

$$\Delta V_{be} = I_1 R$$

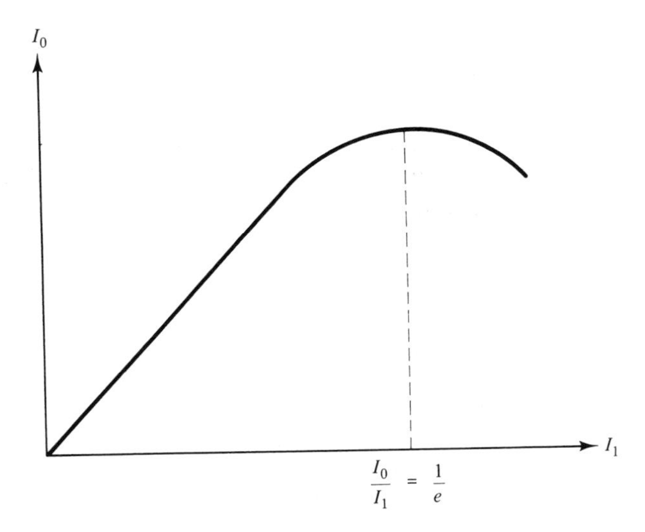

Figure 5-10 Peaking current-source response.

Rearranging the equation and solving for I_0:

$$I_0 = I_1 \exp\left(\frac{-qI_1R}{kT}\right)$$

Taking the partial derivative of I_0 with respect to I_1 yields

$$\frac{\partial I_0}{\partial I_1} = \frac{-qI_1R}{kT} \exp\left(\frac{-qI_1R}{kT}\right) + \exp\left(\frac{-qI_1R}{kT}\right)$$

Setting this equal to zero to locate the maximum:

$$0 = \frac{-qI_1R}{kT} \exp\left(\frac{-qI_1R}{kT}\right) + \exp\left(\frac{-qI_1R}{kT}\right)$$

$$0 = 1 - \frac{qI_1R}{kT}$$

Rearranging:

$$I_1R = \frac{kT}{q}$$

At the peak:

$$I_0 = I_1 \exp(-1)$$

Figure 5-11 illustrates the *npn* and *pnp* implementation of this circuit. If a peaking current source is used to control a bias string, a regulation of $\exp(-1)$ can be achieved about the peak of I_0. This can be used to compensate for current-determining resistor variations with absolute value or temperature. The penalty is that

$$I_1\left(1 - \frac{1}{e}\right)\mu\text{A}$$

of current is "thrown away" to achieve this regulation.

Current sources have, thus far, been described in terms of ideal transistors. Unfortunately, ideal transistors are not offered on most analog ASIC arrays. The next section will describe the effects of finite beta and finite output impedance on the operation of transistor current sources.

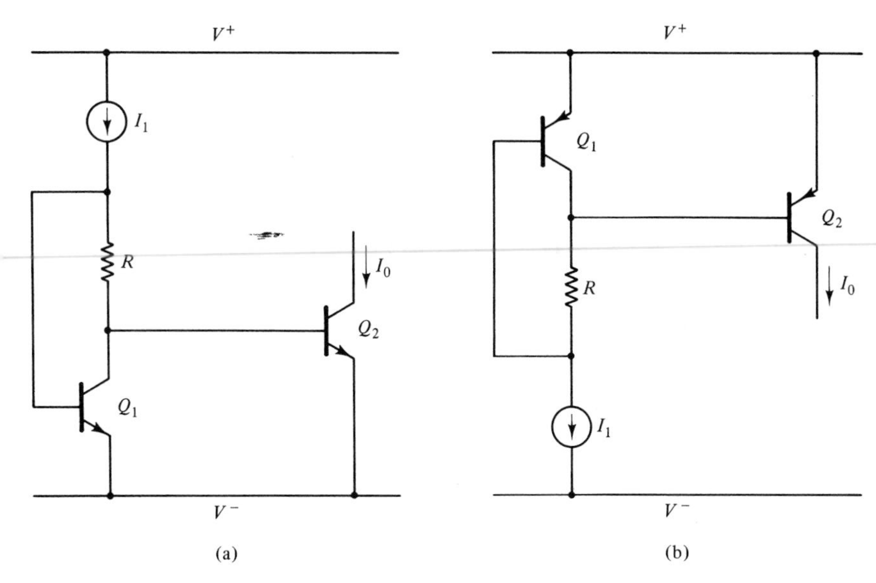

Figure 5-11 Peaking current-source implementations. (*a*) *npn* implementation and (*b*) *pnp* implementation.

Current Source Errors Due to Finite Beta

Figure 5-12 illustrates the error in a current mirror due to finite transistor beta. These simple current mirrors are typically used as collector loads for the input stage of an op amp. This current imbalance has the same effect as an input offset voltage. This will be discussed further in the section on amplifiers. The reference current flowing into the diode side of the mirror must supply the base currents for both transistors. The transfer function for this circuit is[1]

$$I_2 = I_1 - 2I_b$$

$$I_b = \frac{I_2}{\beta}$$

$$I_2 = I_1 - 2\frac{I_2}{\beta}$$

$$I_1 = I_2\left(\frac{\beta + 2}{\beta}\right)$$

$$I_2 = I_1\left(\frac{\beta}{\beta + 2}\right)$$

For beta $= 100$, $I_2 = 0.98I_1$, which represents a 2% error.

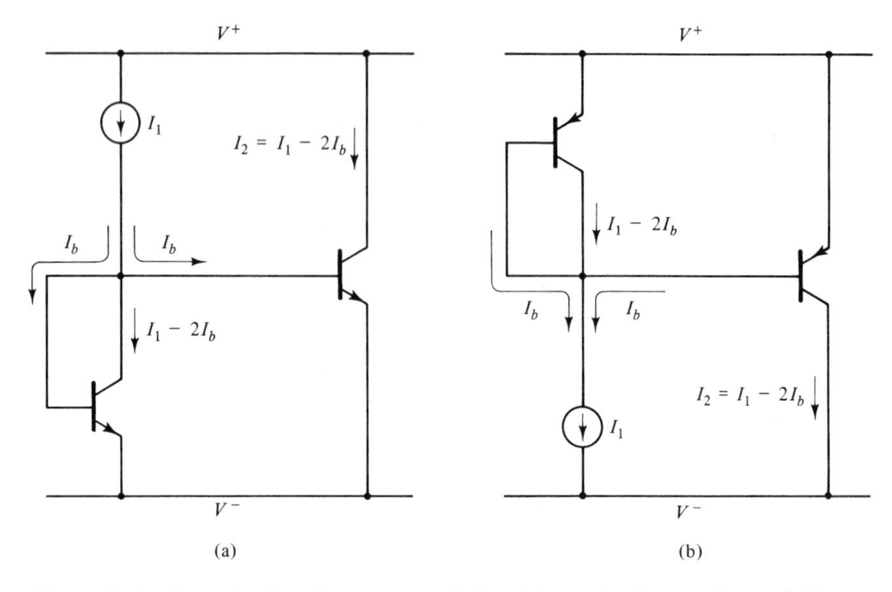

Figure 5-12 Current mirror base current diodes. (*a*) *npn* implementation and (*b*) *pnp* implementation.

On long bias lines with several transistors, the problem is aggravated by the multiple base currents, as shown in Figure 5-13.

Two techniques can be used on the current mirror to significantly reduce this imbalance. The buffer transistor, illustrated in Figure 5-14, supplies the base current for both of the mirror transistors. The imbalance is now the base current of the buffer transistor. This circuit divides the $2I_b$ error current by the beta of the buffer transistor. The transfer function for this circuit is[2,3]

$$I_2 = I_1 - 2\frac{I_b}{\beta}$$

$$I_b = \frac{I_2}{\beta}$$

$$I_2 = I_1 - 2\frac{I1_2}{\beta^2}$$

$$I_2\left(\frac{\beta^2 + 2}{\beta^2}\right) = I_1$$

$$I_2 = I_1\left(\frac{\beta^2}{\beta^2 + 2}\right)$$

For beta $= 100$, $I_2 = 0.9998I_1$, which represents a 0.02% error.

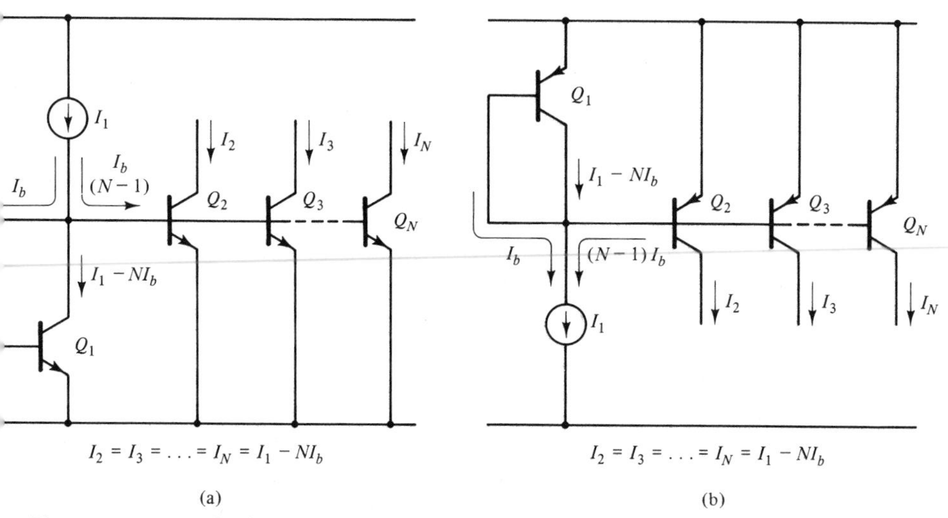

Figure 5-13 Base-buffered errors on bias lines. (*a*) *npn* implementation and (*b*) *pnp* implementation.

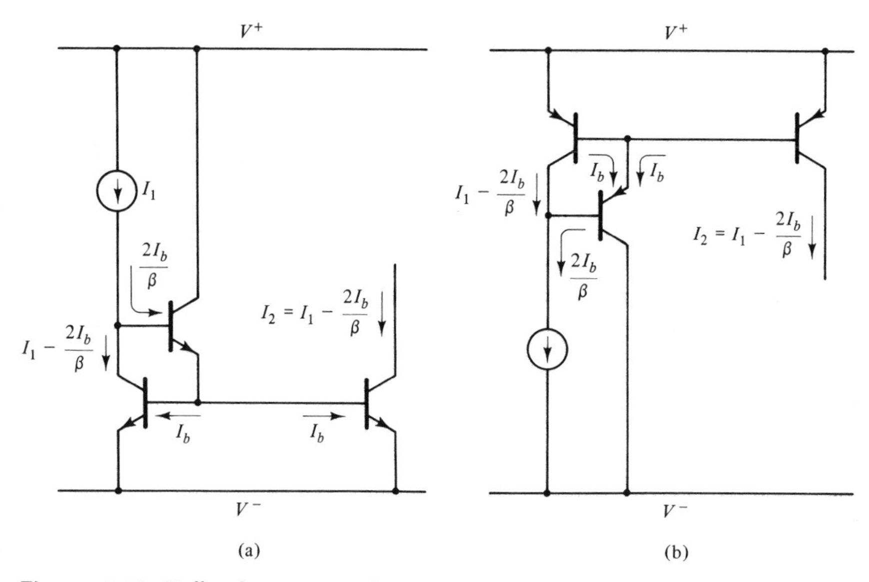

Figure 5-14 Buffered current mirror. (*a*) *npn* implementation and (*b*) *pnp* implementation.

The second technique, frequently called the Wilson current mirror, is illustrated in Figure 5-15. This circuit cancels offsets caused by base current errors. Assuming Q_1, Q_2, and Q_3 are all well matched, the circuit functions as follows: the input source I_1 provides base current I_b to Q_3 and $I_1 - I_b$ to the collector of Q_2. Since Q_1 and Q_2 have the same base-emitter voltage, the collector current of Q_2 will be $I_1 - I_b$. Since $2I_b$ must flow into the bases of Q_1 and Q_2, the emitter current of Q_3 must be $I_1 - I_b + 2I_b$ or $I_1 + I_b$. Since the I_b in Q_3's emitter is its own base current, the collector current of Q_3 is simply I_1. The collector current of Q_3 is I_b greater than that of Q_1 or Q_2 and there will be a mismatch proportional to $1/\beta^2$.[4,5]

There are a few limitations with this circuit. The collector voltage of Q_2 is clamped at $V_{be}Q_1 + V_{be}Q_3$ and the collector voltage of Q_3 can go no lower than $V_{be}Q_1 + V_{sat}Q_3$. Although this circuit provides greater accuracy, it limits the voltage compliance of the current mirror. This can be of concern in low power-supply voltage applications and in circuits where common mode voltage range is important.

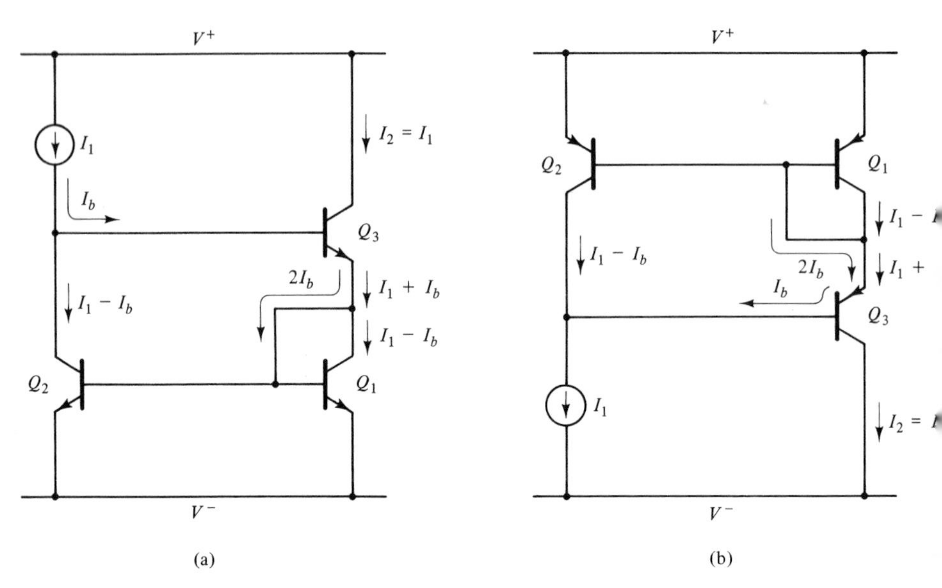

Figure 5-15 Wilson current mirror. (*a*) *npn* implementation and (*b*) *pnp* implementation.

Current Source Errors Due to Finite Output Impedance

In addition to beta errors, the output impedance of the current source transistor can make a significant contribution to current matching errors. For demonstration, a simple current mirror, illustrated in Figure 5-16, will be analyzed. It will be assumed that the extrapolated Early voltage, V_a, is 200 V for the *npn* transistors (Figure 5-16(*a*)) and 100 V for the *pnp* transistors (Figure 5-16(*b*)). It will be further assumed that the collector of Q_2 is at a potential of 10 V in the *npn* case and ground in the *pnp* case. The output impedance of Q_2 can be calculated as follows:[6]

$$r_0 = \frac{V_a}{I_c}$$

For Figure 5-16(*a*) the *npn* current mirror

$$r_0 = \frac{200\ \text{V}}{40\ \mu\text{A}} = 5\ \text{M}\Omega$$

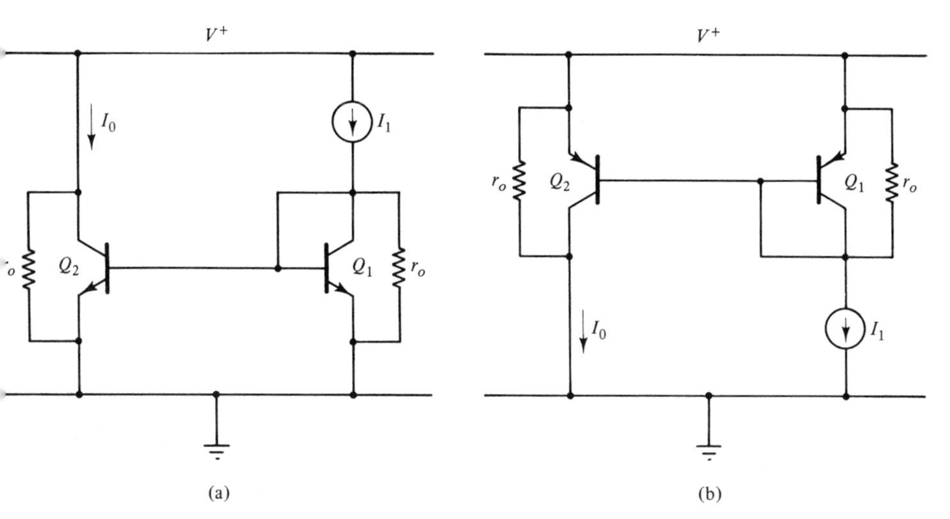

Figure 5-16 Output errors on current matching. (*a*) *npn* current mirror and (*b*) *pnp* current mirror.

For Q_1:

$$I = \frac{10\ \text{V}}{5\ \text{M}\Omega} = 2\ \mu\text{A}$$

For Q_2:

$$I = \frac{0.65\ \text{V}}{5\ \text{M}\Omega} = 0.13\ \mu\text{A}$$

The mismatch is $\approx 1.9\ \mu$A

$$\text{mismatch} = \frac{1.9\ \mu\text{A}}{40\ \mu\text{A}} \times 100\% \approx 5\%$$

For Figure 5-16(*b*), the *pnp* current mirror

$$r_0 = \frac{100\text{V}}{40\ \mu\text{A}} = 2.5\ \text{M}\Omega$$

For Q_1:

$$I = \frac{10\ \text{V}}{2.5\ \text{M}\Omega} = 4\ \mu\text{A}$$

For Q_2:

$$I = \frac{0.65\ \text{V}}{2.5\ \text{M}\Omega} = 0.26\ \mu\text{A}$$

The mismatch is $\approx 3.8\ \mu$A

$$\text{mismatch} = \frac{3.8\ \mu\text{A}}{40\mu\text{A}} \times 100\% \approx 9.5\%$$

This error is much worse for the *pnp* due to its lower effective r_0. This, combined with the inherently lower *pnp* beta, means that realizing precision circuitry is more difficult with *pnp* transistors than with *npn* transistors.

Keep in mind that the mismatch due to r_0 is a function of the collector voltage on Q_2. If this current mirror is used in the input stage of an op amp as the collector load, the error would be a function of the amplifier's common-mode voltage.

As with beta errors, there are circuit implementations that can be used to mitigate output impedance related problems. The first technique is a modified Wilson current source. The *npn* and *pnp* realizations of this circuit are illustrated in Figure 5-17(*a*) and 5-17(*b*). In these circuits the output impedance of Q_2 and Q_3 track and the output impedance of the diode-connected transistors Q_1 and Q_4 track. The net result is that the output impedance errors are effectively cancelled to the degree that the transistors and currents through them match.

The major disadvantages of this circuit are the complexity and the minimum voltage ($2V_{be} \approx 1.2$ V) the circuit requires to operate. As the complexity of the current source increases it is more difficult to lay out. All of the transistors in the Wilson current source must match and track very closely. This means they must be located close together on the die and have the same orientation. It is important to note that, even if a circuit is designed using this configuration, the potential performance improvements over a simple current mirror depend on the layout of the circuit on the die. If the transistors must be separated in order to be connected in this configuration, thermal gradients could eliminate any

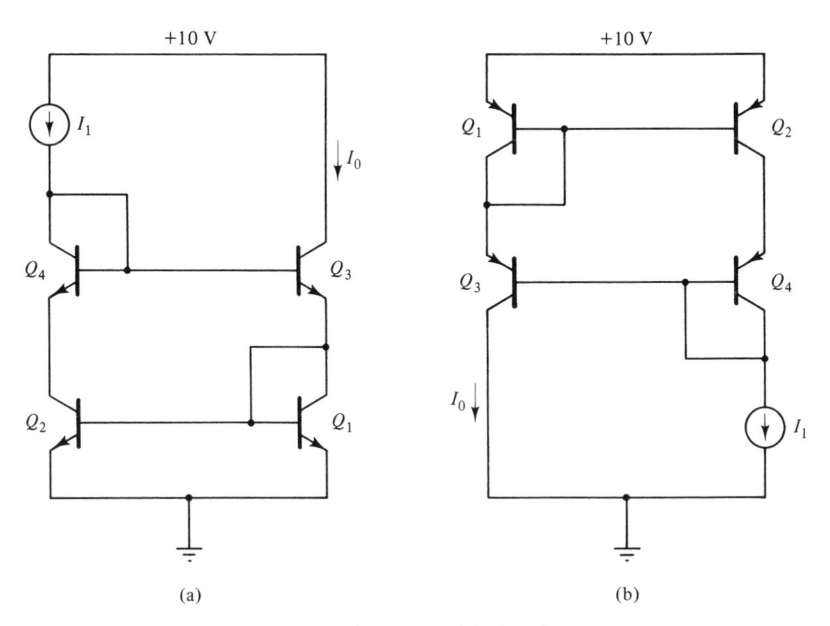

Figure 5-17 (*a*) *npn* and (*b*) *pnp* modified Wilson current sources.

potential advantage or actually degrade its performance to the point that it is not as good as a simple current mirror.

A very high output impedance current source can be realized using the "cascode" arrangement illustrated in Figure 5-18.[7] The term cascode is an acronym left over from vacuum tube design and stands for "cascaded cathode to anode." This circuit places the output impedance of Q_2 and Q_4 in series. Q_3 simply provides proper base bias for Q_4 and determines the collector voltage of Q_2. The circuits shown in Figure 5-18 have Q_2 biased with a collector-base voltage of 0 V. This provides for the maximum voltage compliance for output current I_0. Q_3 has no effect on the magnitude of I_0 and merely provides voltage bias. Q_3 can be replaced with two or more diode-connected transistors, a resistor, or any other voltage reference. The effect will be to modulate the collector voltage on Q_2.

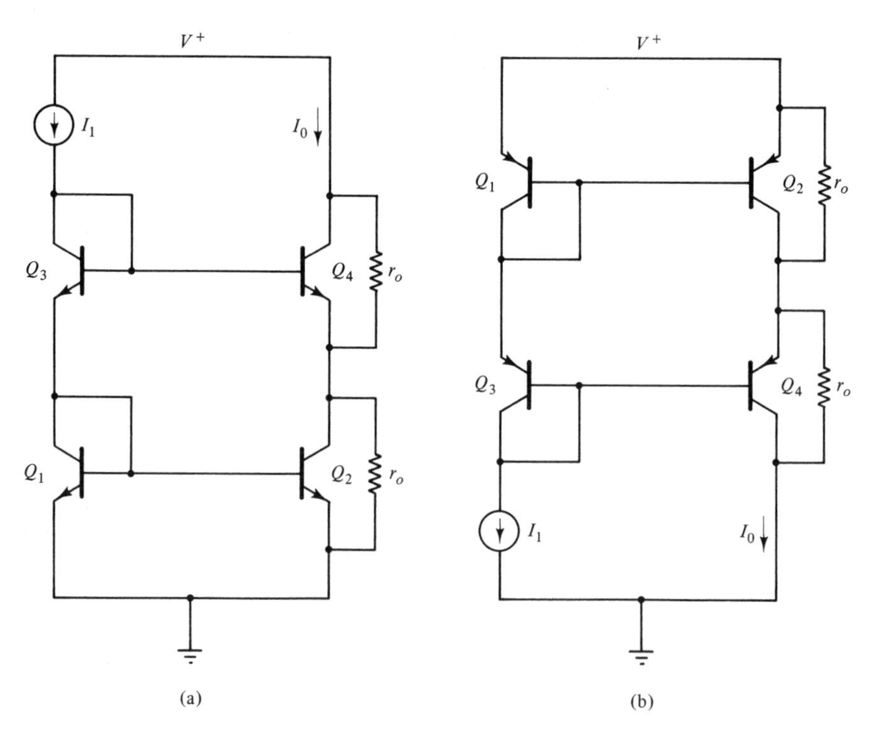

Figure 5-18 Cascode transistors improve output impedance. (*a*) *npn* and (*b*) *pnp* cascode arrangements.

5.2 Voltage References

A voltage reference can be used for many different purposes in a circuit. It can be a rock steady reference with a known absolute value and a very low temperature coefficient used as a reference for A/D or D/A conversion or for biasing of the circuitry on a chip. A voltage reference can be created with a known temperature coefficient or resistor value dependence to cancel another effect. When operating on battery power, it is frequently desirable for circuit bias voltages to be proportional to the battery voltage to prolong the usable battery life. For example, amplifier circuits are frequently biased such that the output is set at one-half the battery voltage. This way, as the battery discharges, the center of the output swing remains at one-half of the battery voltage. Although the maximum signal swing may decrease as the battery discharges, the swing above and below the reference point will remain symmetrical.

There are many different implementations of voltage references. Although many are closely related, they can be roughly broken into four major groups; resistive dividers, diode, Zener, and band gap. Resistor dividers are the simplest and most widely used to provide reference voltages proportional to other voltages. Resistor dividers can use current sources and transistors as well as resistors. The other three types of voltage references provide some measure of regulation and tend to output a constant voltage even if the input voltage changes.

Resistive Dividers

The most basic resistor divider voltage reference is illustrated in Figure 5-19. This circuit can generate any desired reference voltage less than V, depending on available resistor values. The resistor values available on analog semicustom arrays are fixed. The values are often in integer ratios. To obtain other than integer ratios, resistors will have to be connected in series and/or in parallel to achieve the desired value. Remember that the values of resistors in series add and the equivalent value of resistors in parallel is determined by the following relationship:

$$R_{eq} = \frac{1}{\dfrac{1}{R_1} + \dfrac{1}{R_2} + \dfrac{1}{R_3} + \ldots + \dfrac{1}{R_n}}$$

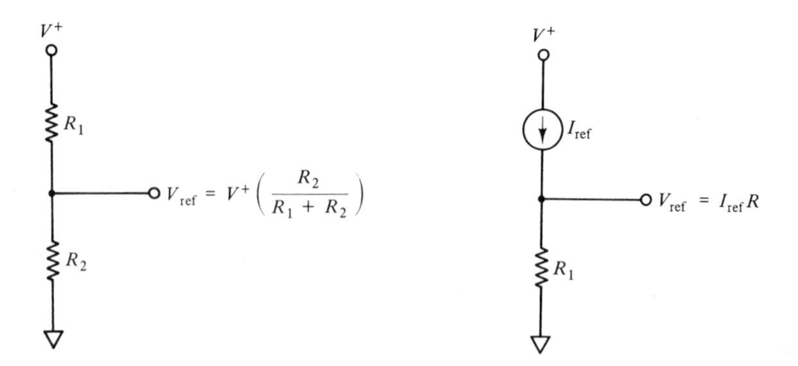

Figure 5-19 Voltage divider reference.

Figure 5-20 Resistor-current source reference.

For two resistors in parallel:

$$R_{eq} = \frac{R_1R_2}{R_1 + R_2}$$

For three resistors in parallel

$$R_{eq} = \frac{R_1R_2R_3}{R_2R_3 + R_1R_3 + R_1R_2}$$

Always be careful of the loading on the reference terminal of a resistive divider. If this is not considered, the initial reference can be different than expected. Changes in loading resulting from temperature changes or process variations on a unit to unit basis can also impact the value of V_{ref}.

Figure 5-20 illustrates an alternate approach for using a resistor to generate a reference voltage. The current can be generated any number of ways to achieve the desired voltage and temperature coefficient.

Figure 5-21 shows one possible implementation of a resistor/current source reference. Techniques described in Section 5.1 on current sources such as transistor area ratios and emitter ballast resistors can be used in addition to resistor combinations to achieve nearly any desired reference voltage. Loading on the reference node can affect the value of V_{ref}.

Other resistive divider reference circuits use transistors for buffering to reduce the effects of loading the reference node and diodes to add a temperature coefficient or cancel the temperature coefficient of the buffer transistor's V_{be}. Figures 5-22 through 5-25 illustrate various

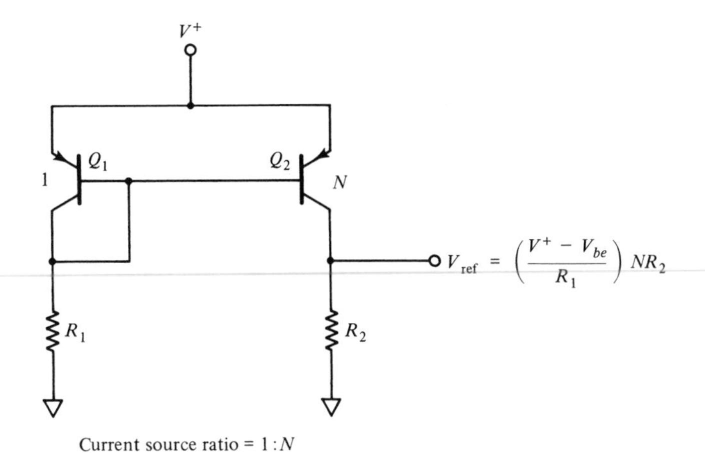

Figure 5-21 Ratioed resistor-current source reference.

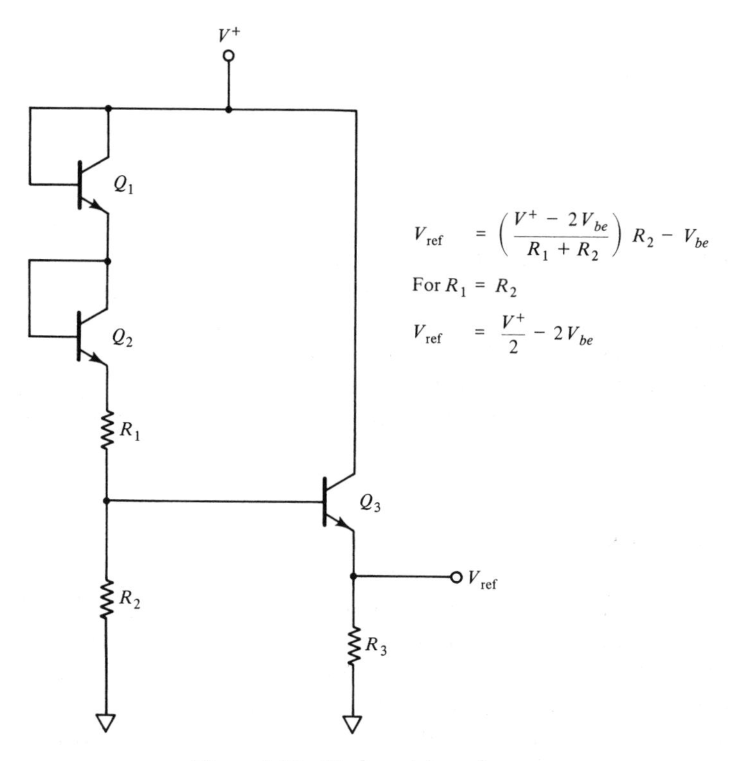

Figure 5-22 Diode-resistor reference.

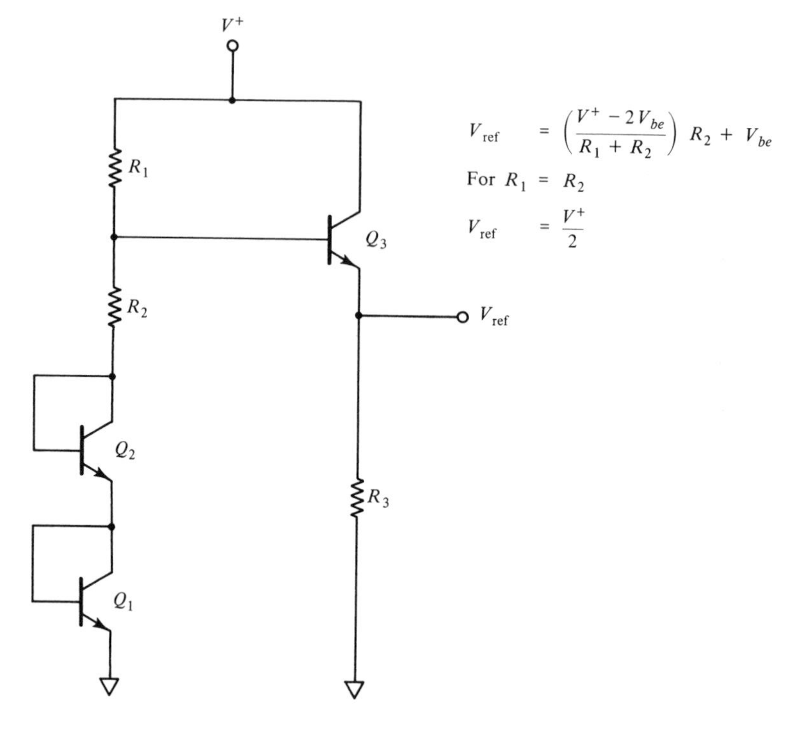

Figure 5-23 Diode-resistor reference.

implementations of these references. These are not the only possible implementations—many variations are possible—but serve to demonstrate some basic concepts. The equations for the reference voltage assume that the effects of the buffer transistor's base current are negligible. Be sure this assumption is valid in any particular application using one of these circuits. The temperature coefficient on a V_{be} is approximately -2.2 mV/°C.

Diode References

Diode references depend on the voltage of a forward-biased base-emitter junction. A typical value for a V_{be} is 0.65 V. This voltage depends on semiconductor process, temperature, and current. A typical V_{be} has a temperature of approximately -2.2 mV/°C and a voltage coefficient of approximately $+60$ mV/decade increase in current (this does not include resistive effects).

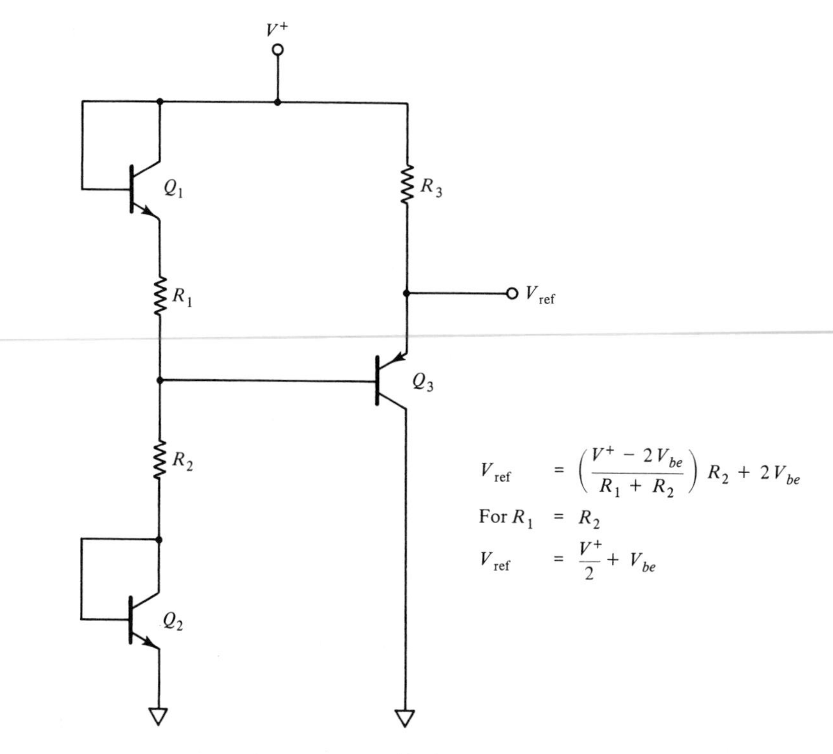

Figure 5-24 Diode-resistor reference.

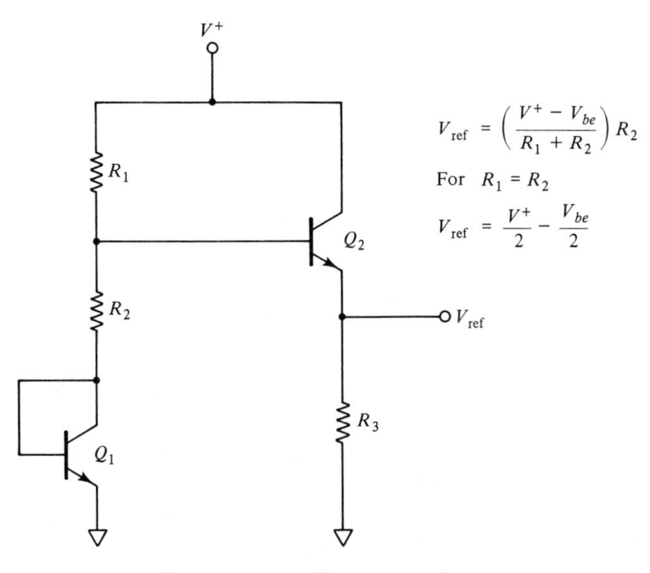

Figure 5-25 Diode-resistor reference.

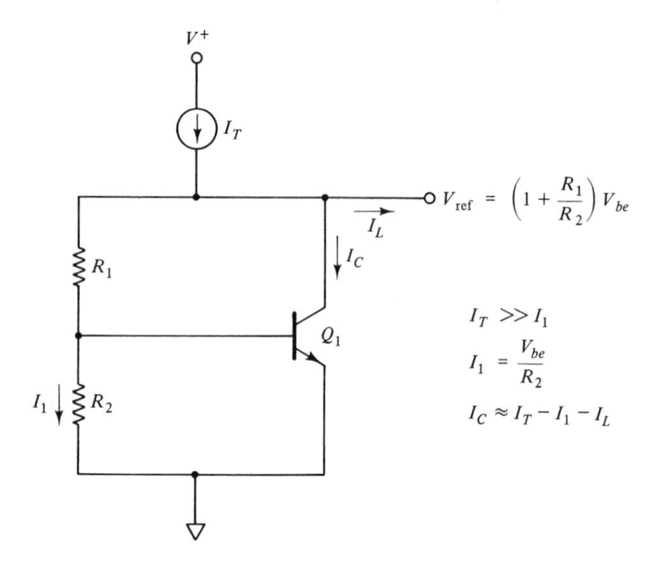

Figure 5-26 V_{be} multiplier.

Figure 5-26 illustrates a V_{be} multiplier circuit.[8,9] For this circuit to operate properly, I_T must be larger than $I_1 + I_L$. I_1 develops a V_{be} across R_2 thus turning on Q_1. The voltage across R_1 is

$$V_{R1} = \frac{V_{be}R_1}{R_2}$$

The reference voltage is the combination of the voltage drop across R_1 and R_2.

$$V_{ref} = V_{R1} + V_{R2} = V_{be} + V_{be}\frac{R_1}{R_2}$$

$$V_{ref} = V_{be}\left(1 + \frac{R_1}{R_2}\right)$$

The difference between I_1 and I_L is shunted by Q_1, thus providing load regulation.

Figure 5-27 illustrates a method of generating a buffered 1 V_{be} reference. This circuit also generates, in the collector of Q_1, a current equal to V_{be}/R. Current source I supplies base current for Q_3 and bias current for the two diodes. The voltage

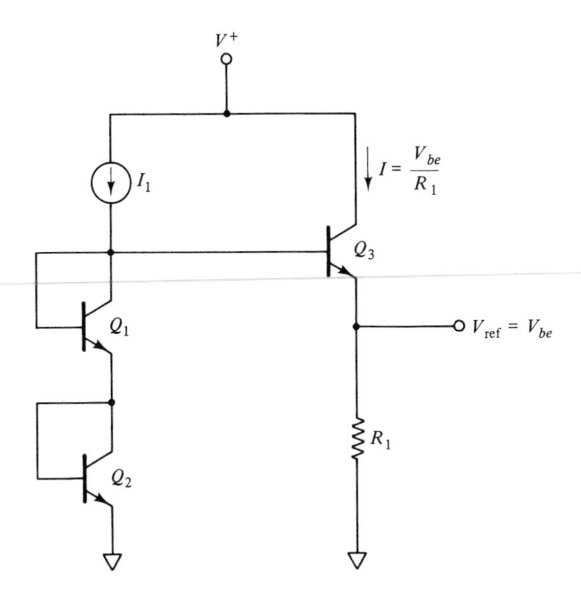

Figure 5-27 (V_{be}/R) reference.

$$V_{ref} = V_{be_1} + V_{be_3} - V_{be_3}$$

appears across the resistor R. This circuit can be extended, as shown in Figures 5-28 to generate a reference voltage of

$$V_{ref} = (N - 1)V_{be}$$

and a current of

$$I = \frac{NV_{be}}{R}$$

Each V_{be} added to V_{ref} increases the temperature coefficient of V_{ref} by -2.2 mV/°C. The temperature coefficient of Q_3's collector current is similarly affected.

Zener References

The circuit in Figure 5-29 illustrates a simple Zener reference. Typical emitter base breakdown voltages are approximately 6.3 V and have a

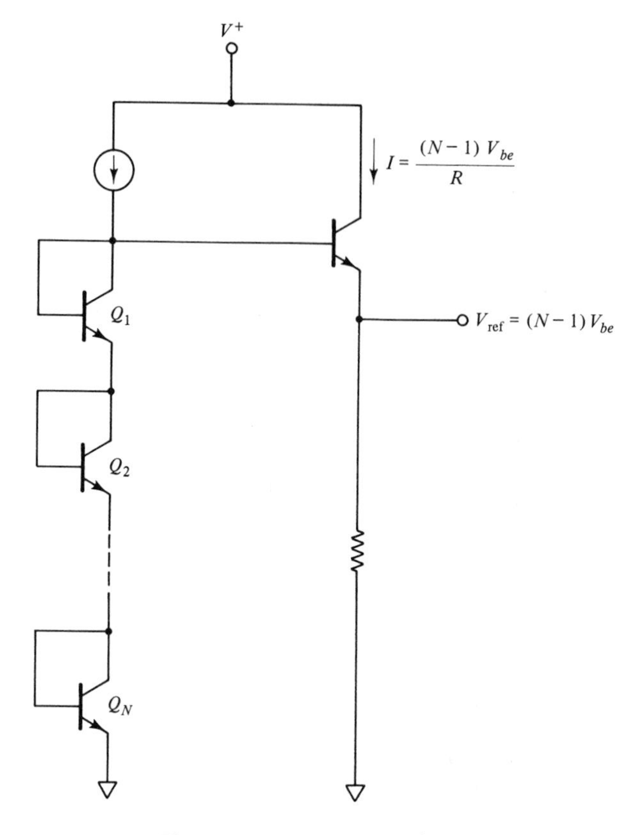

Figure 5-28 $(N - 1)V_{be}$ reference.

temperature coefficient of approximately +2.5 mV/°C. Both the Zener voltage and its temperature coefficient depend somewhat on the semiconductor process and bias current. The temperature coefficient of V_{ref} can be reduced by adding the forward-biased diode Q_2, shown in Figure 5-29(b). The reference voltage will now be

$$V_{ref} = V_z + V_{be} \approx 6.3\text{ V} + 0.65\text{ V} \approx 6.95\text{ V}$$

The temperature coefficient will be

$$\text{T.C.} = +2.5\text{ mV/°C} - 2.2\text{ mV/°C} = +0.3\text{ mV/°C}$$

Variations of this circuit are shown in Figures 5-30 and 5-31.

The V_{be} multiplier circuit, discussed previously, can be used with Zeners to provide temperature compensation, a wide range of voltages,

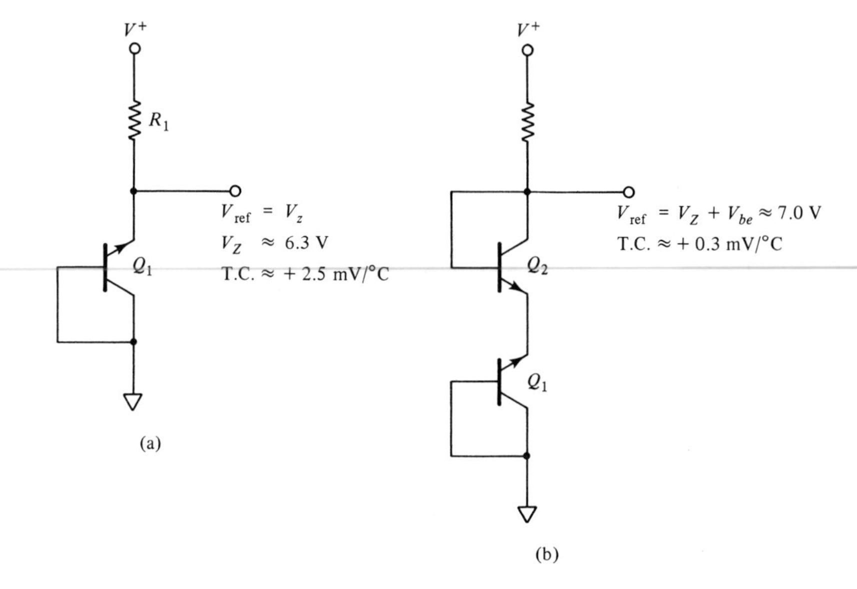

Figure 5-29 Simple Zener references.

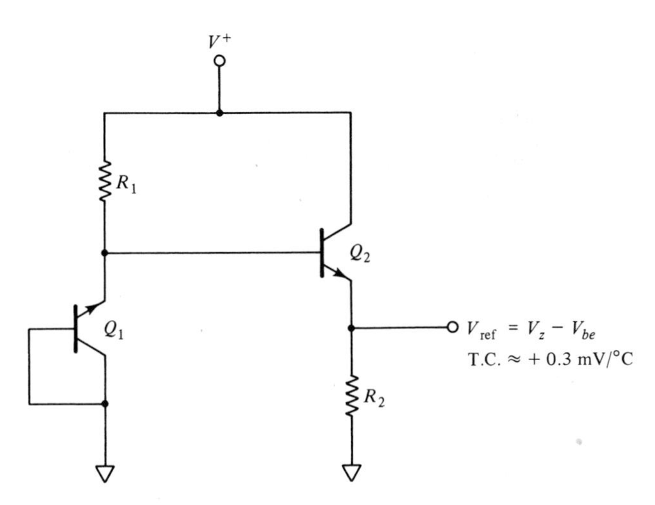

Figure 5-30 Buffered $V_z - V_{be}$ reference.

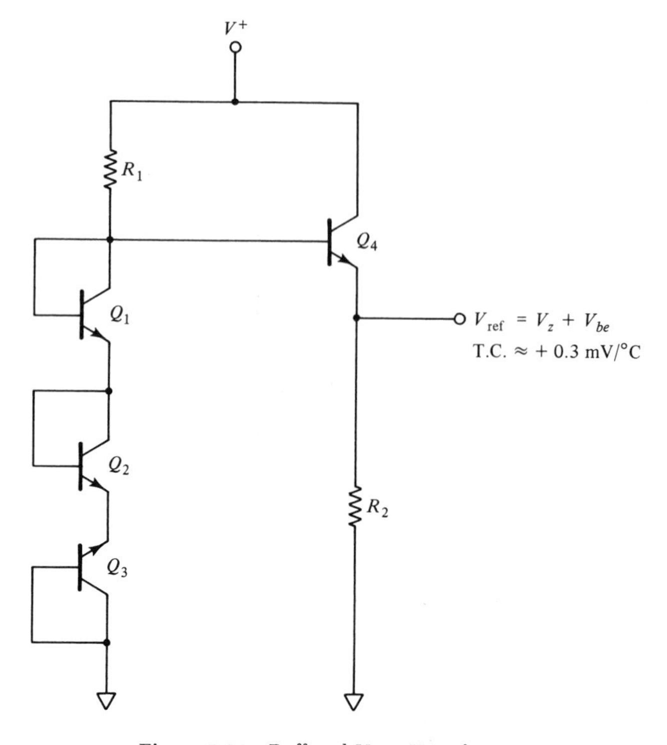

Figure 5-31 Buffered $V_z + V_{be}$ reference.

and shunt load regulation. Figure 5-32 shows a Zener used in the basic configuration. I_T must be larger than $I_1 + I_L$.

$$V_{ref} = V_z + V_{be}$$

The temperature coefficient is approximately

$$+2.5\ \text{mV/°C} - 2.2\ \text{mV/°C} \approx +0.3\ \text{mV/°C}$$

Figure 5-33 illustrates a variation of the basic circuit described above that enables the temperature coefficient of V_{ref} to be adjusted to zero with the appropriate selection of R_1. By selecting larger values of R_1, the value of V_{ref} can be increased but will have a negative temperature coefficient.

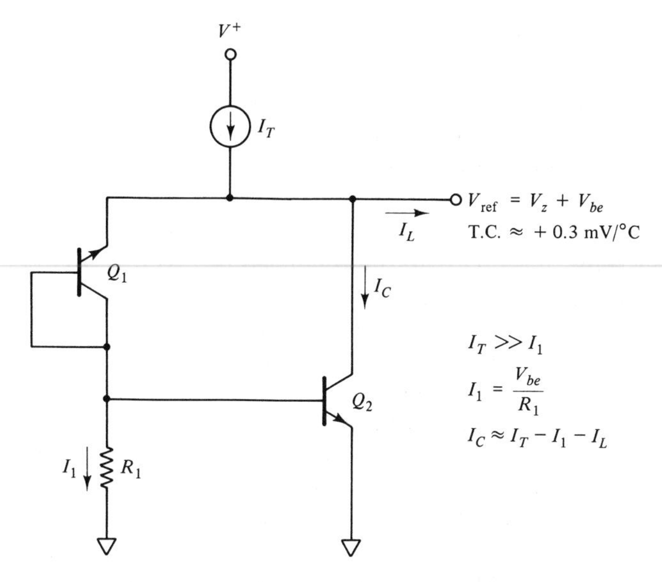

Figure 5-32 Shunt-regulated Zener reference.

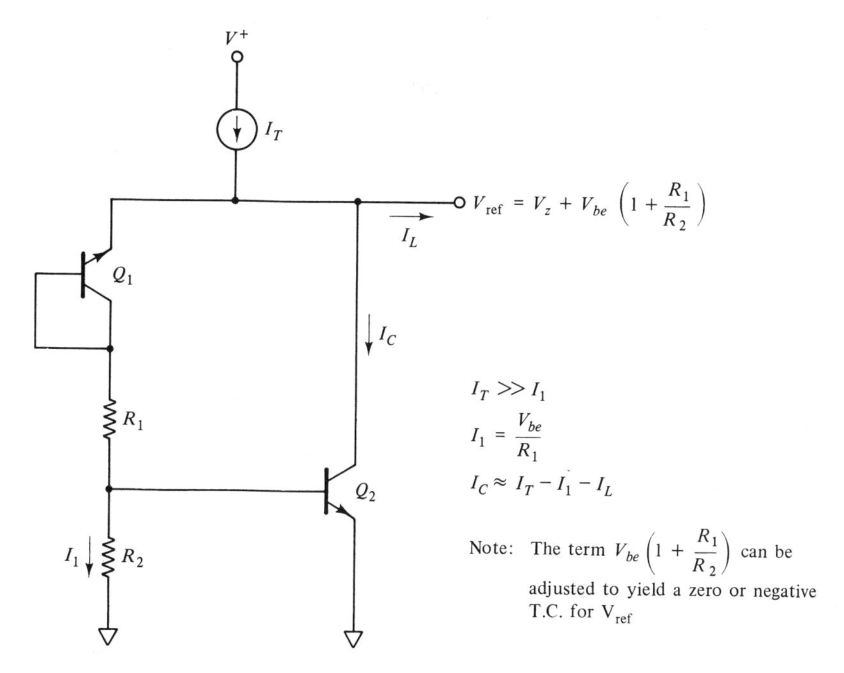

Figure 5-33 Temperature-compensated Zener reference. (Note: the term $V_{be}[1 + (R_1/R_2)]$ can be adjusted to yield a zero or negative T.C. for V_{ref}.

Bandgap References

All of the references discussed thus far are relatively simple but have drawbacks, depending on the desired precision. Resistive references depend on matching and tracking of resistors and a reference voltage (the power supply), and then introduce thermal noise. Diode references are imprecise in both absolute value and temperature coefficient. Zener references require a relatively high voltage ($\approx$7 to 8 V) to operate and introduce a significant amount of noise.

There is another type of reference that is more complex but offers performance advantages over those previously discussed. These are referred to as bandgap references.[10,11] They derive their name from the fact that the output voltage of the reference is nearly equal to the bandgap voltage of silicon. The exact zero temperature coefficient output voltage depends on semiconductor process characteristics such as doping concentrations and device geometries. The discussion here will be general in nature to explain the basic concept. For more details consult references and your ASIC vendor. Many ASIC vendors offer bandgap references as predesigned, precharacterized cells. For simplicity and predictable results, use these whenever possible.

Bandgap references function by balancing the negative temperature coefficient of a V_{be} with the positive temperature coefficient of a difference in two V_{be}s. The V_{be} difference is generated by subtracting the V_{be}s of two transistors operating at different current densities. The current density difference is most often realized by using transistors with different emitter areas.

Widlar Reference

The first implementation, commonly referred to as the Widlar bandgap reference, is illustrated in Figure 5-34. The V_{be} difference is generated by the N:1 area ratio of Q_1 and Q_2. Q_2 has a lower V_{be} than Q_1 since it has N times larger emitter area. The difference in V_{be}s is developed across R_3. The resulting current flows through Q_2 and R_2. The ratio of R_2 and R_3 multiplies the difference in V_{be} which appears across R_2. The multiplied V_{be} difference appearing on R_2, which has a positive temperature coefficient, is added to the V_{be} of Q_3, which has a negative temperature coefficient, to yield V_{ref}. It can be shown mathematically that the reference will have a zero temperature coefficient when V_{ref}

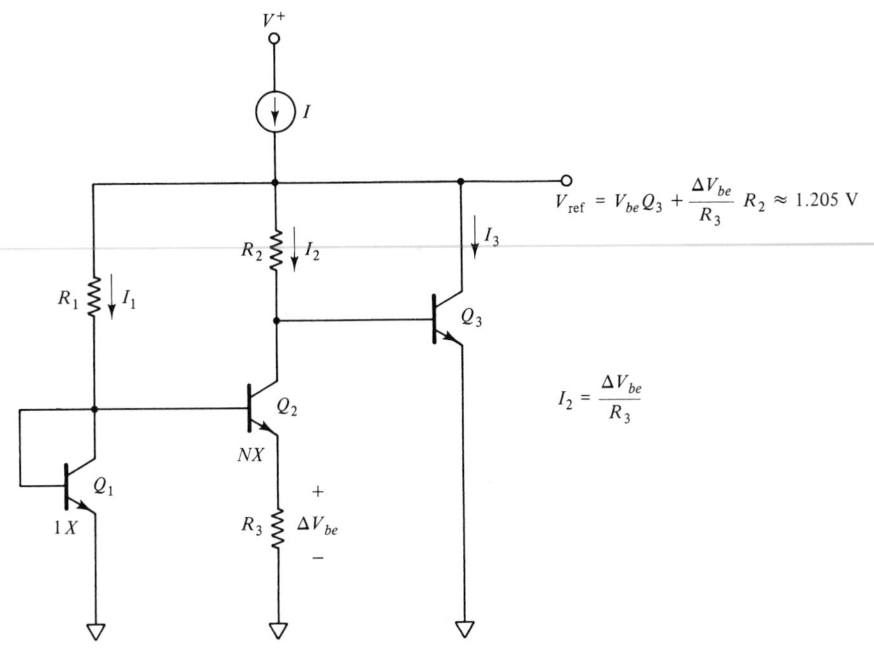

Figure 5-34 Widlar bandgap reference.

is set equal to the bandgap of silicon, which is approximately 1.205 V. In addition to providing a V_{be}, Q_3 provides shunt regulation of the current I.

Brokaw Reference

The second implementation of the bandgap reference functions similarly to the first but is a more versatile configuration. Figure 5-35 illustrates the basic circuit. The current through Q_1 and Q_2 is the same due to the symmetrical current mirror Q_3 and Q_4. The V_{be} of Q_2 is less than that of Q_1 since the emitter area of Q_2 is N times larger than Q_1. The difference in V_{be} is developed across R_3 and the resulting current from Q_1 and Q_2 flows through R_2. As before, the ratio between R_2 and R_3 determines the multiplication of the V_{be} difference term. The multiplied V_{be} difference appears across R_2 and is added to the V_{be} of Q_1. The 0 T.C. point occurs when the bases of Q_1 and Q_2 are at a potential equal

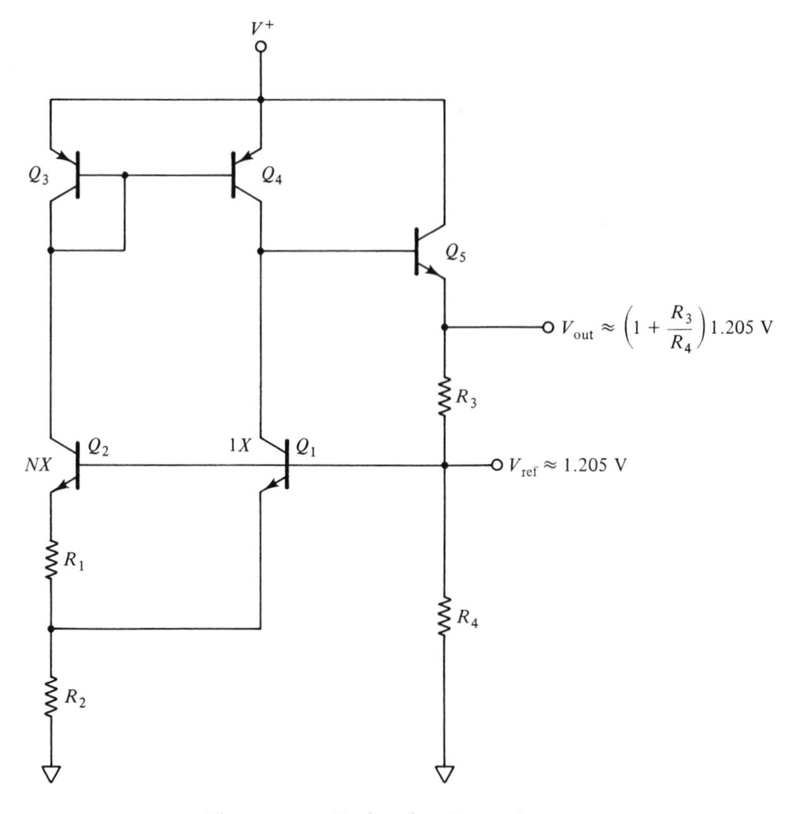

Figure 5-35 Brokaw bandgap reference.

to the bandgap voltage of silicon. Q_4 serves as a high impedance collector load for Q_1. The resulting loop gain, buffering action of Q_5 and the resistor divider (R_4, R_5) can be used to set V_{out} to any desirable voltage. The output voltage V_{out} is determined by the following relationship:

$$V_{out} \left(\frac{R_4 + R_5}{R_5}\right)(1.205 \text{ V})$$

The previous discussions about the two different bandgap references neglected some higher order effects. First, base currents were assumed to be negligible. More importantly, the bandgap voltage of silicon was implied to be constant. It is not. The bandgap potential of silicon changes as temperature is varied above and below 300 K. This variation affects the stability of the references previously described. Additional circuitry can be added to provide correction for this effect and thereby a reference with improved temperature stability.

If a highly stable reference is required for your application, the use of a commercially available reference should be strongly considered. These circuits are extremely sophisticated and would be difficult to duplicate on an ASIC array with fixed transistor geometries, fixed resistor values, layout constraints, and with other circuitry and the attendant thermal gradients on the same chip.

5.3 Amplifiers

Analog circuitry represents or models physical characteristics such as temperature, pressure, mechanical vibration, and light level as electrical equivalents or analogs of the characteristic in question. This electrical equivalent is typically in the form of a voltage or current that varies in a linear correspondence to the physical quantity being modeled. Due to the nature of these physical quantities and the available transducers to convert them into voltages or currents, the resulting electrical signals can be very small. To be useful, these signals must be amplified or have characteristics such as spectral content modified. Amplifiers are therefore central to any analog signal processing.

Amplifiers can be simple or relatively complex, depending on the desired level of performance and ease of use. Simple one-, two-, or three-stage resistively biased amplifiers are adequate in many applications. They typically require capacitive coupling, have a relatively low input impedance, and low gain. Improving the gain of this type of amplifier usually requires the use of bypass capacitors. The use of capacitors has three major impacts on the integration of circuitry. Capacitors are physically large components. They cannot be integrated and take up valuable circuit board area. They seriously impact the low-frequency response of the amplifier. Resistively biased amplifiers can be open loop with a fixed gain or fed back like an op amp.

Figure 5-36 illustrates a simple resistively biased single-transistor amplifier. R_1, R_2 and R_E provide emitter bias while R_C serves as the collector load. C_1 and C_2 isolate the input source and load, respectively, from the dc potentials appearing on the base and collector nodes of the transistor. R_S is the source impedance and R_L is the load impedance.

Figure 5-37(*a*) shows the equivalent input circuit of the amplifier in Figure 5-36, and Figure 5-37(*b*) shows the equivalent output circuit. The only differences between the *npn* and *pnp* implementations of this

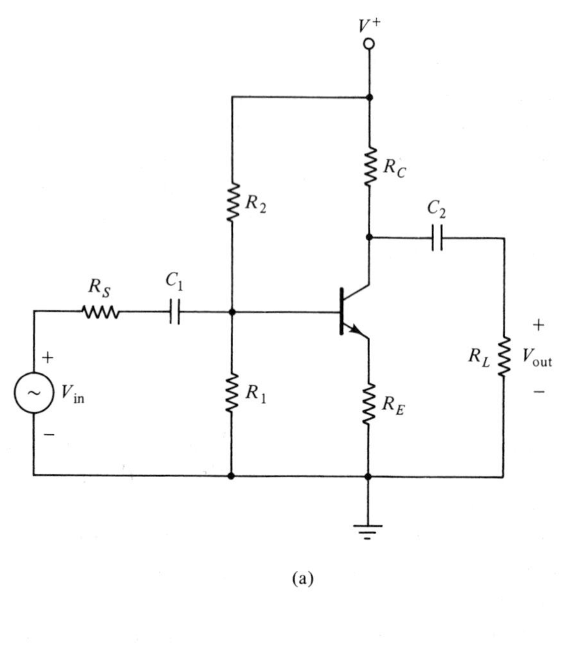

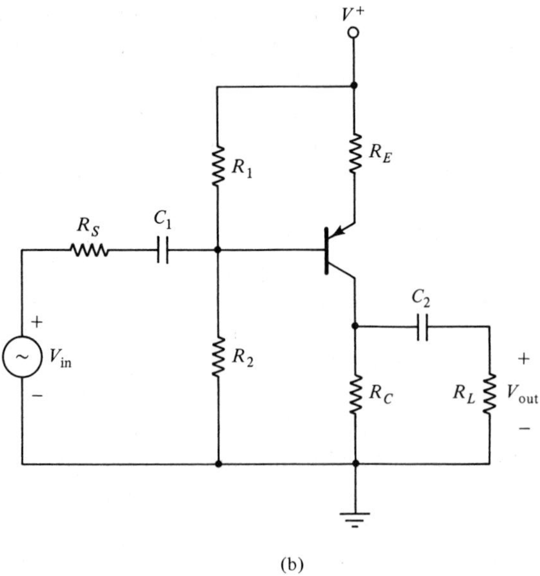

Figure 5-36 Single-transistor, resistively biased amplifier. (*a*) *npn* implementation and (*b*) *pnp* implementation.

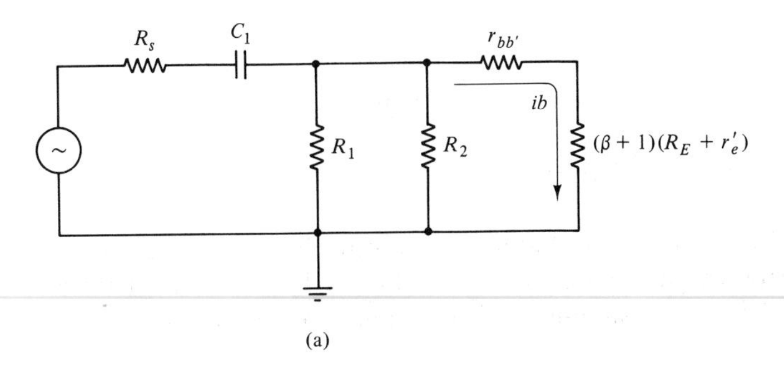

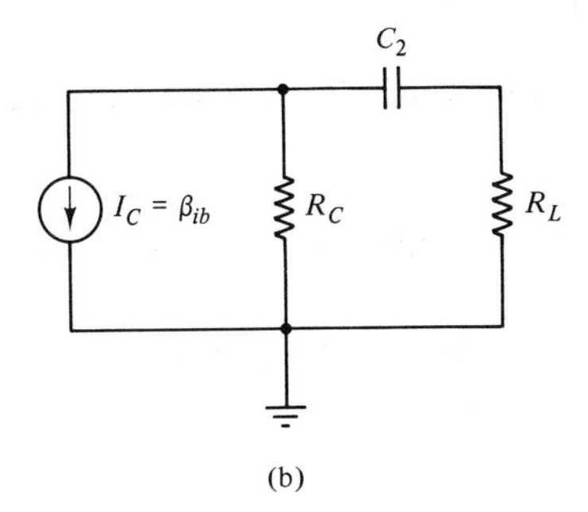

Figure 5-37 Equivalent low-frequency (*a*) input and (*b*) output circuits.

circuit are the interchanging of V^+ and ground and the reversal of current flow. These two implementations are equivalent if the reduced performance of the *pnp* transistor, such as beta and low current handling ability, is accounted for. The input circuit forms a high-pass filter with a corner frequency equal to

$$f = \frac{1}{2\pi R_{eq} C_1}$$

where

$$R_{eq} = R_S // R_1 // R_2 // (r_{bb'} + (\beta + 1)(R_e + r_e))$$

The output circuit also forms a high-pass circuit with a corner frequency equal to

$$f = \frac{1}{2\pi R_{eq} C_2}$$

where

$$R_{eq} = R_C // R_L$$

Low-Frequency Amplifier

As an example, we will analyze the circuit in Figure 5-38. We will assume that the frequencies are well below f_β of the transistor and therefore the reactive elements of the transistor model can be neglected.[12]

Calculate the dc Bias Point

Figure 5-39*a* shows the circuit for calculating the base voltage. Assumptions:

1. The impedance looking into the base of Q_1 does not load the base bias circuit.
2. The base current is negligible compared to the current through R_1 and R_2.
3. $V_{be} = 0.6$ V

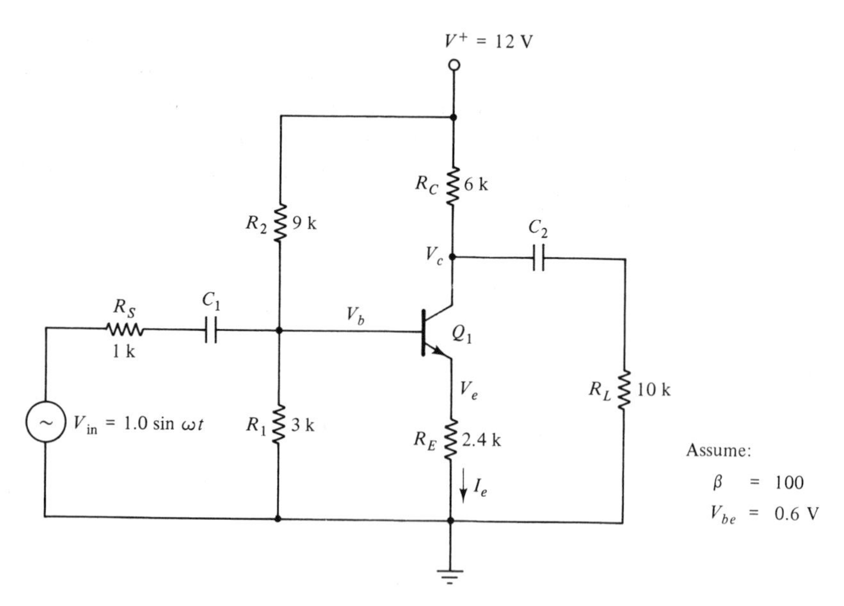

Figure 5-38 Resistively biased amplifier. (Assume $\beta = 100$ and $V_{be} = 0.6$ V.)

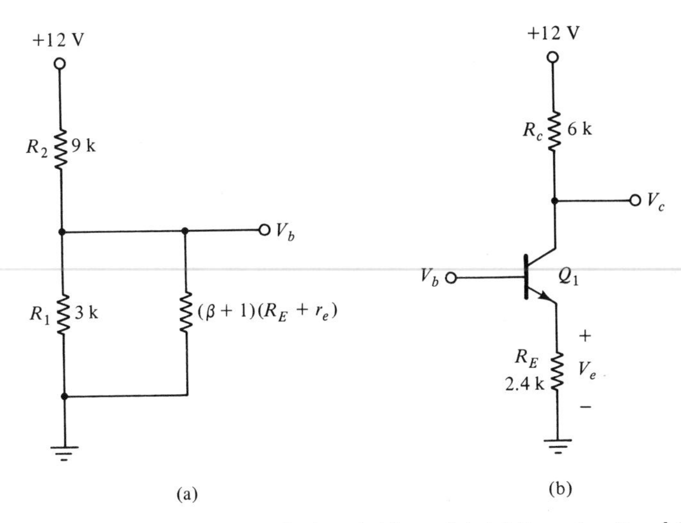

Figure 5-39 Equivalent circuit to calculate dc bias point. (*a*) Base circuit and (*b*) output circuit.

$$V_b = \frac{(12\ \text{V})(R_1)}{R_2 + R_1} = \frac{(12)(3\ \text{k}\Omega)}{12\ \text{k}\Omega}\text{V} = 3\ \text{V}$$

$$V_e = V_b - 0.6\ \text{V} = 2.4\ \text{V}$$

$$I_e = \frac{2.4\ \text{V}}{2.4\ \text{k}\Omega} = 1.0\ \text{mA}$$

$$I_c = \frac{\beta}{\beta + 1} I_e = \frac{100}{101} 1.0\ \text{mA} \approx 1.0\ \text{mA}$$

$$V_c = 12\ \text{V} - I_c R_c = 12\ \text{V} - (1.0\ \text{mA})(6\ \text{k}\Omega) = 6\ \text{V}$$

Calculate Gain

Figure 5-40 shows the equivalent circuit for calculating ac gain. Assumptions:

1. The impedance looking into the base of Q_1 does not load the base bias circuit.
2. The signal frequency is well above the corner frequency due to C_1 and C_2.

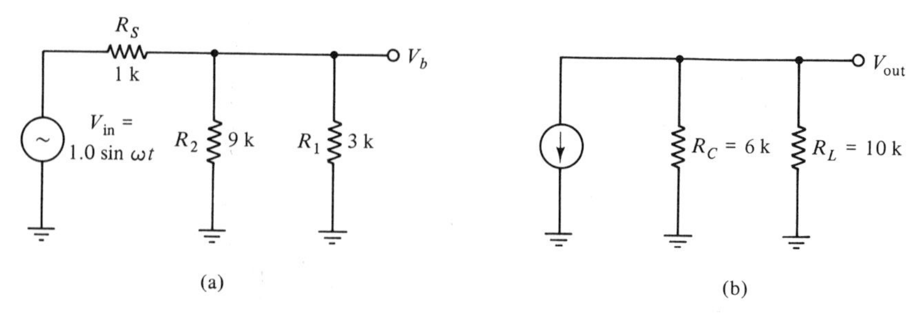

Figure 5-40 Equivalent circuit to calculate ac gain. (*a*) Base circuit and (*b*) collector circuit.

$$V_b = \frac{(V_{in})(R_2//R_1)}{R_S + R_2//R_1}$$

$$R_2//R_1 = \frac{(9\ \text{k}\Omega)(3\ \text{k}\Omega)}{9\ \text{k}\Omega + 3\ \text{k}\Omega} = 2.25\ \text{k}\Omega$$

$$V_b = \frac{2.25\ \text{k}\Omega}{3.25\ \text{k}\Omega}(1.0 \sin \omega t) = 0.692 \sin \omega t$$

$$V_e \approx 0.692 \sin \omega t$$

$$i_e \approx \frac{0.692 \sin \omega t}{2.4\ \text{k}\Omega} = 0.288 \sin \omega t\ \text{mA} \approx i_c$$

$$V_{out} \approx i_c(R_C//R_L)$$

$$R_C//R_L = \frac{R_C R_L}{R_C + R_L} = \frac{(6\ \text{k}\Omega)(10\ \text{k}\Omega)}{6\ \text{k}\Omega + 10\ \text{k}\Omega} = 3.75\ \text{k}\Omega$$

$$V_{out} \approx (0.288\ \text{mA})(3.75\ \text{k}\Omega)(\sin \omega t)$$

$$V_{out} \approx 1.08 \sin \omega t$$

Next, choose values for C_1 and C_2 such that the corner frequency is 500 Hz.

Calculate C_1

$$C_1 = \frac{1}{2\pi f R_{eq}}$$

$$R_{eq} = R_S//R_1//R_2//(\beta + 1)(R_E + r_e)$$

Assuming the impedance looking into the base of Q_1 contributes negligible loading,

$$R_{eq} = 2.25 \text{ k}\Omega // 1 \text{ k}\Omega$$

$$R_{eq} = \frac{(2.25 \text{ k}\Omega)(1 \text{ k}\Omega)}{3.25 \text{ k}\Omega} = 692.3 \ \Omega$$

$$C_1 = \frac{1}{2\pi(500 \text{ Hz})(692.3 \ \Omega)} = 0.46 \ \mu\text{F}$$

***Calculate* C_2**

$$C_2 = \frac{1}{2\pi f R_{eq}}$$

$$R_{eq} = R_L // R_C$$

$$R_{eq} = 10 \text{ k}\Omega // 6 \text{ k}\Omega$$

$$R_{eq} = 3.75 \text{ k}\Omega$$

$$C_1 = \frac{1}{2\pi(500 \text{ Hz})(3.75 \text{ k}\Omega)} = 0.085 \ \mu\text{F}$$

To check the hand analysis of the circuit we will run a dc and ac SPICE analysis. The circuit to be analyzed is shown in Figure 5-41 and includes node numbers.

The SPICE input file is shown in Figure 5-42 and the output file in Figure 5-43. The dc bias points and gain calculated by SPICE are slightly different from the hand calculations due to the simplifying assumptions made for the hand analysis. Figure 5-44 shows the frequency and phase response of the amplifier.

This example has served several purposes. First, several of the basic limitations of simple resistively biased amplifiers have been illustrated. Low input impedance causes a loading effect on the driving source (previous stage). Gain is inherently low due to R_E and a relatively low value of R_C. The load impedance R_L causes a voltage division with R_C, which reduces achievable gain. The bias currents are relatively high. Coupling capacitors are necessary due to the dc voltages at the input and output of the amplifier. This type of amplifier is difficult to use, at low frequencies or dc. Cascaded poles in an amplifier (or several stages of these amplifiers) cause the bandwidth to change. Sec-

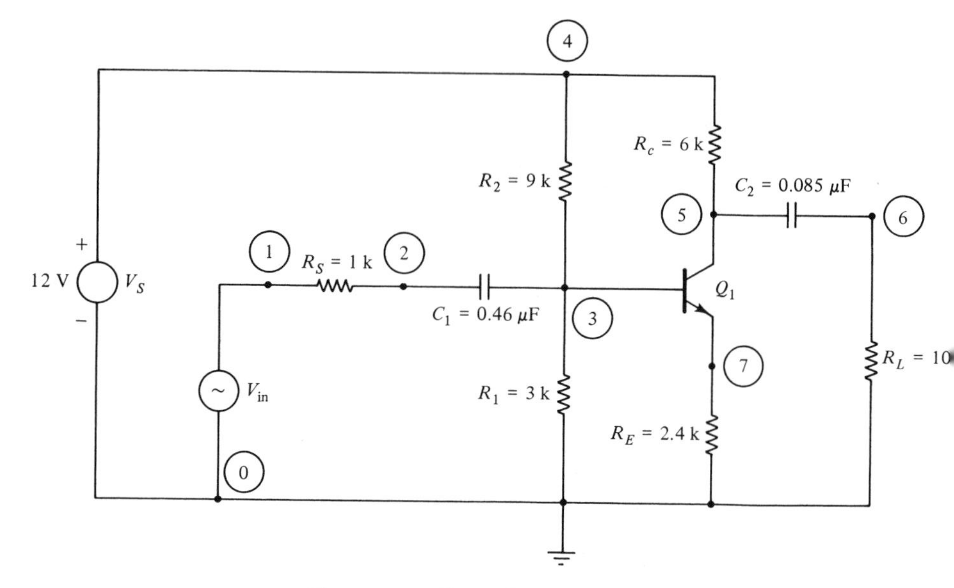

Figure 5-41 Circuit for SPICE analysis.

ond, analyzing this circuit gave us the opportunity to do a SPICE analysis. Finally, understanding the limitations of this circuit will reveal various ways to circumvent them.

The circuit in Figure 5-45 illustrates three relatively simple additions to the basic circuit that significantly reduces several of its major shortcomings. The addition of the resistor R_3 increases the input impedance of the amplifier by isolating the input from the (R_1, R_2) voltage divider. The emitter bypass capacitor (C_E) increases the achievable ac gain by providing a low impedance for ac signals on Q_1's emitter. The addition of Q_2 buffers the load from R_C, thus eliminating the previously observed gain reduction.

A more easily integrated implementation of this circuit is illustrated in Figure 5-46. The use of split supplies (±6 V), current source biasing, and a differential input stage eliminates the need for coupling capacitors. The gain of the amplifier depends on R_C and r_e of Q_1 and Q_2, which is determined by the emitter bias current. In other words, the gain of the amplifier is controlled by R_{bias} (or the current in the diode Q_4). By adding four transistors to the circuit we have eliminated three capacitors which cannot be integrated and made the circuit usable on dc signals.

```
LOW FREQUENCY AMPLIFIER                BY: PAUL BROWN

***CIRCUIT DESCRIPTION***
*This is a resistively biased single transistor amplifier
*that is capacitively coupled to both the source and load.
*
*
***CIRCUIT CONNECTION***

***RESISTORS***

R1   0   3   3K
R2   3   4   9K
RC   5   4   6K
RE   0   7   2.4K
RL   0   6   10K
RS   1   2   1K
***CAPACITORS***

C1   2   3   0.46UF
C2   5   6   0.085UF
***SEMICONDUCTORS***

Q1   5   3   7   SGNPN
***SOURCES***

VS   4   0   12V
VIN  1   0   AC   1V
***MODEL DEFINITION***

.MODEL   SGNPN   NPN IS=5E-16 BF=150 VAF=200
+IKF=1E-3 ISE=1E-13 ISC=1E-12 RB=150 RE=2 RC=50
+CJE=2.5E-12 VJE=0.6 TF=0.33E-9 CJC=2E-12 TR=10E-9
+CJS=2E-12 XTB=2
***ANALYSIS CONTROL***
.AC DEC 10 10 1MEG
***OUTPUT CONTROL***
.PRINT AC V(1) VP(1) VDB(6) VP(6)
.END
```

Figure 5-42 Low-frequency amplifier SPICE input file.

The circuit in Figure 5-46 is very similar to what is called an operational amplifier or op amp. Before we develop this subject further it is important to review the high-frequency performance of a single-transistor amplifier and then look at an integrated implementation of a high-frequency differential amplifier (sometimes called a video amplifier). This will also prepare us to look in depth at the high-performance limitations of op amps and other types of circuitry such as multipliers, to be discussed later.

```
******* 2/ 3/91 ******* IS SPICE v1.5M 4/14/89 *******11:51:59*****

LOW FREQUENCY AMPLIFIER          BY: PAUL BROWN

****     INPUT LISTING                    TEMPERATURE =   27.000 DEG C

***********************************************************************



***CIRCUIT DESCRIPTION***
*This is a resistively biased single transistor amplifier *that is capacitively
*
*
***CIRCUIT CONNECTION***

***RESISTORS***

R1      0       3       3K
R2      3       4       9K
RC      5       4       6K
RE      0       7       2.4K
RL      0       6       10K
RS      1       2       1K
***CAPACITORS***

C1      2       3       0.46UF
C2      5       6       0.085UF
***SEMICONDUCTORS***

Q1      5       3       7       SGNPN
***SOURCES***

VS      4       0       12V
VIN     1       0       AC      1V
***MODEL DEFINITION***

.MODEL  SGNPN   NPN IS=5E-16 BF=150 VAF=200 +IKF=1E-3 ISE=1E-13 ISC=1E-12 RB=150
***ANALYSIS CONTROL***
.AC DEC 10 10 1MEG
***OUTPUT CONTROL***
.PRINT AC V(1) VP(1) VDB(6) VP(6)
.END
******* 2/ 3/91 ******* IS SPICE v1.5M 4/14/89 *******11:51:59*****

LOW FREQUENCY AMPLIFIER          BY: PAUL BROWN

****     BJT MODEL PARAMETERS             TEMPERATURE =   27.000 DEG C

***********************************************************************



             SGNPN
TYPE         NPN
IS          5.00D-16
BF           150.000
NF             1.000
VAF         2.00D+02
ISE         1.00D-13
```

Figure 5-43 Low-frequency amplifier SPICE output file. (*Figure continues.*)

```
BR             1.000
NR             1.000
ISC         1.00D-12
RB           150.000
******* 2/ 3/91 ******* IS SPICE v1.5M 4/14/89 *******11:51:59*****

LOW FREQUENCY AMPLIFIER          BY: PAUL BROWN

****      SMALL SIGNAL BIAS SOLUTION       TEMPERATURE =   27.000 DEG C

************************************************************************

 NODE   VOLTAGE      NODE   VOLTAGE      NODE   VOLTAGE      NODE   VOLTAGE


(  1)     .0000     (  2)     .0000     (  3)    2.9537     (  4)   12.0000

(  5)    6.5710     (  6)     .0000     (  7)    2.2210




    VOLTAGE SOURCE CURRENTS

    NAME         CURRENT


    VS         -1.910D-03

    VIN         0.000D-01


    TOTAL POWER DISSIPATION   2.29D-02  WATTS
******* 2/ 3/91 ******* IS SPICE v1.5M 4/14/89 *******11:51:59*****

LOW FREQUENCY AMPLIFIER          BY: PAUL BROWN

****      OPERATING POINT INFORMATION      TEMPERATURE =   27.000 DEG C

************************************************************************



**** BIPOLAR JUNCTION TRANSISTORS


           Q1
MODEL      SGNPN
IB         2.06E-05
IC         9.05E-04
VBE            .733
VBC          -3.617
VCE           4.350
BETADC       43.927
GM         3.50E-02
RPI        1.65E+03
RX         1.50E+02
RO         2.25E+05
CPI        0.00E-01
CMU        0.00E-01
CBX        0.00E-01
```

Figure 5-43 (*Continued*)

```
CCS        0.00E-01
BETAAC       57.598
FT         5.57E+17
******* 2/ 3/91 ******* IS SPICE v1.5M 4/14/89 *******11:51:59*****

LOW FREQUENCY AMPLIFIER          BY: PAUL BROWN

****     AC ANALYSIS                       TEMPERATURE =   27.000 DEG C

**********************************************************************
```

FREQ	V(1)	VP(1)	VDB(6)	VP(6)
1.00000E+01	1.000E+00	0.000E-01	-4.171E+01	-1.019E+01
1.25893E+01	1.000E+00	0.000E-01	-3.775E+01	-1.281E+01
1.58489E+01	1.000E+00	0.000E-01	-3.381E+01	-1.608E+01
1.99526E+01	1.000E+00	0.000E-01	-2.991E+01	-2.017E+01
2.51189E+01	1.000E+00	0.000E-01	-2.606E+01	-2.524E+01
3.16228E+01	1.000E+00	0.000E-01	-2.230E+01	-3.148E+01
3.98107E+01	1.000E+00	0.000E-01	-1.867E+01	-3.907E+01
5.01187E+01	1.000E+00	0.000E-01	-1.522E+01	-4.814E+01
6.30957E+01	1.000E+00	0.000E-01	-1.203E+01	-5.870E+01
7.94328E+01	1.000E+00	0.000E-01	-9.169E+00	-7.058E+01
1.00000E+02	1.000E+00	0.000E-01	-6.718E+00	-8.339E+01
1.25893E+02	1.000E+00	0.000E-01	-4.717E+00	-9.655E+01
1.58489E+02	1.000E+00	0.000E-01	-3.166E+00	-1.094E+02
1.99526E+02	1.000E+00	0.000E-01	-2.022E+00	-1.213E+02
2.51189E+02	1.000E+00	0.000E-01	-1.213E+00	-1.318E+02
3.16228E+02	1.000E+00	0.000E-01	-6.615E-01	-1.409E+02
3.98107E+02	1.000E+00	0.000E-01	-2.942E-01	-1.485E+02
5.01187E+02	1.000E+00	0.000E-01	-5.419E-02	-1.547E+02
6.30957E+02	1.000E+00	0.000E-01	1.008E-01	-1.598E+02
7.94328E+02	1.000E+00	0.000E-01	2.000E-01	-1.639E+02
1.00000E+03	1.000E+00	0.000E-01	2.632E-01	-1.672E+02
1.25893E+03	1.000E+00	0.000E-01	3.033E-01	-1.698E+02
1.58489E+03	1.000E+00	0.000E-01	3.287E-01	-1.719E+02
1.99526E+03	1.000E+00	0.000E-01	3.448E-01	-1.736E+02
2.51189E+03	1.000E+00	0.000E-01	3.550E-01	-1.749E+02
3.16228E+03	1.000E+00	0.000E-01	3.614E-01	-1.759E+02
3.98107E+03	1.000E+00	0.000E-01	3.654E-01	-1.768E+02
5.01187E+03	1.000E+00	0.000E-01	3.680E-01	-1.774E+02
6.30957E+03	1.000E+00	0.000E-01	3.696E-01	-1.780E+02
7.94328E+03	1.000E+00	0.000E-01	3.706E-01	-1.784E+02
1.00000E+04	1.000E+00	0.000E-01	3.712E-01	-1.787E+02
1.25893E+04	1.000E+00	0.000E-01	3.716E-01	-1.790E+02
1.58489E+04	1.000E+00	0.000E-01	3.719E-01	-1.792E+02
1.99526E+04	1.000E+00	0.000E-01	3.721E-01	-1.794E+02
2.51189E+04	1.000E+00	0.000E-01	3.722E-01	-1.795E+02
3.16228E+04	1.000E+00	0.000E-01	3.722E-01	-1.796E+02
3.98107E+04	1.000E+00	0.000E-01	3.723E-01	-1.797E+02
5.01187E+04	1.000E+00	0.000E-01	3.723E-01	-1.797E+02
6.30957E+04	1.000E+00	0.000E-01	3.723E-01	-1.798E+02
7.94328E+04	1.000E+00	0.000E-01	3.723E-01	-1.798E+02
1.00000E+05	1.000E+00	0.000E-01	3.723E-01	-1.799E+02
1.25893E+05	1.000E+00	0.000E-01	3.723E-01	-1.799E+02
1.58489E+05	1.000E+00	0.000E-01	3.723E-01	-1.799E+02
1.99526E+05	1.000E+00	0.000E-01	3.723E-01	-1.799E+02
2.51189E+05	1.000E+00	0.000E-01	3.723E-01	-1.799E+02
3.16228E+05	1.000E+00	0.000E-01	3.723E-01	-1.800E+02
3.98107E+05	1.000E+00	0.000E-01	3.723E-01	-1.800E+02
5.01187E+05	1.000E+00	0.000E-01	3.723E-01	-1.800E+02
6.30957E+05	1.000E+00	0.000E-01	3.723E-01	-1.800E+02
7.94328E+05	1.000E+00	0.000E-01	3.723E-01	-1.800E+02
1.00000E+06	1.000E+00	0.000E-01	3.723E-01	-1.800E+02

```
        JOB CONCLUDED
         TOTAL JOB TIME             4.10
```

Figure 5-43 *(Continued)*

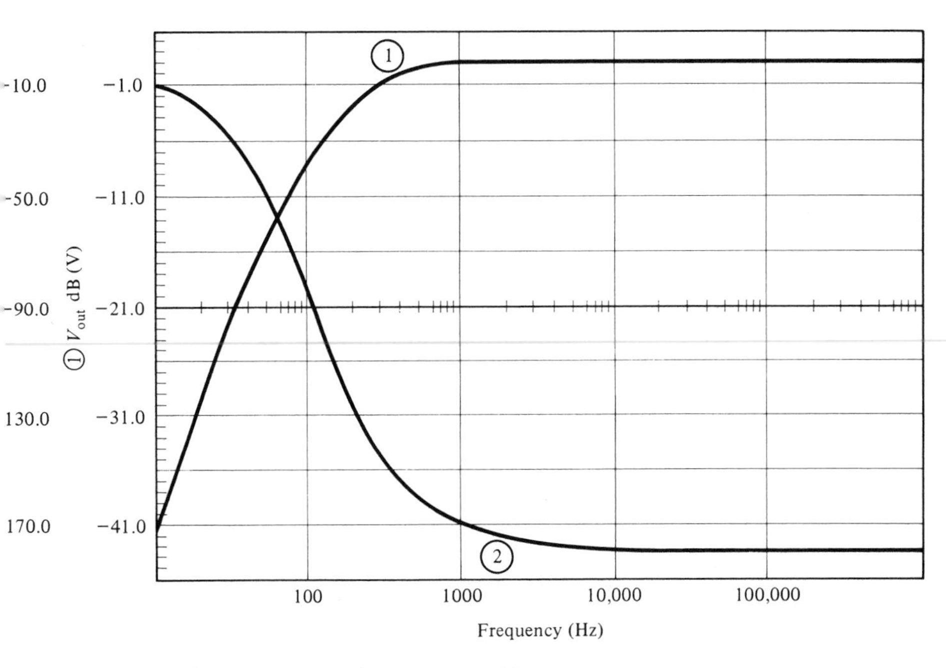

Figure 5-44 Low-frequency amplifier ac output response.

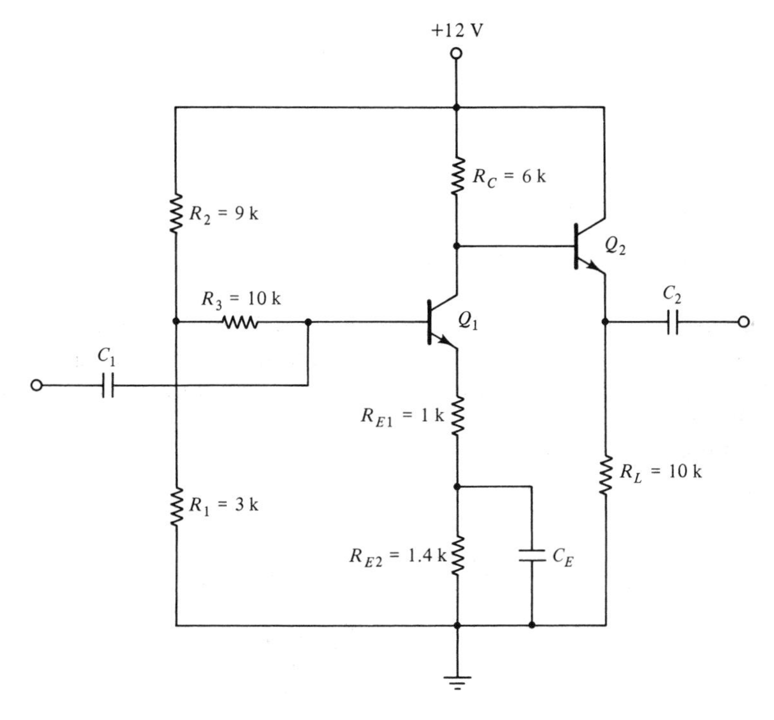

Figure 5-45 Improved resistively biased amplifier.

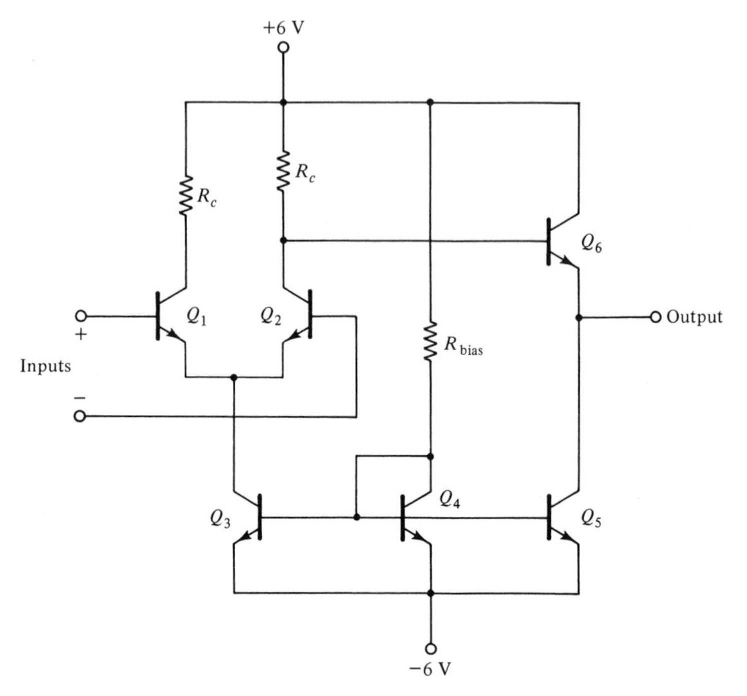

Figure 5-46 An easily integratable amplifier.

High-Frequency Transistor Model

The high-frequency hybrid-π model for a transistor is illustrated in Figure 5-47.[13] Figure 5.47(*a*) shows the collector-base capacitance as $C_{b'c}$. An equivalent Miller capacitance can be reflected into the base circuit as follows:

The voltage across $C_{b'c}$ is

$$V_{C_{b'c}} = V_{b'e} - V_{ce}$$

$$V_{ce} = \frac{-V_{b'e}\beta R_L}{r_{b'e}} = -\frac{V_{b'e}(\beta)}{(\beta + 1)r_e} R_L$$

$$V_{C_{b'e}} = V_{b'e}\left(1 + \frac{\beta}{\beta + 1}\frac{R_L}{r_e}\right)$$

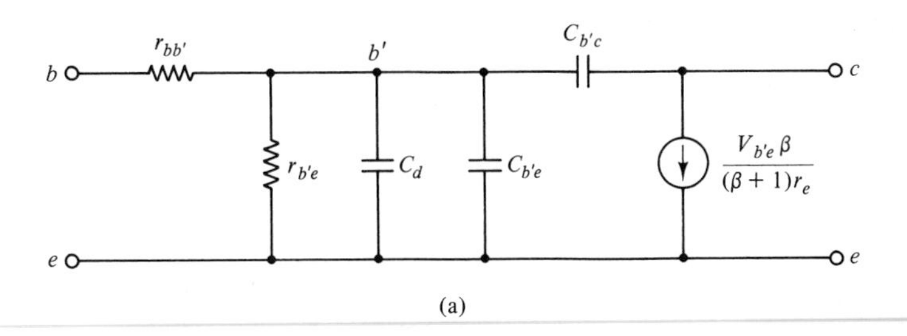

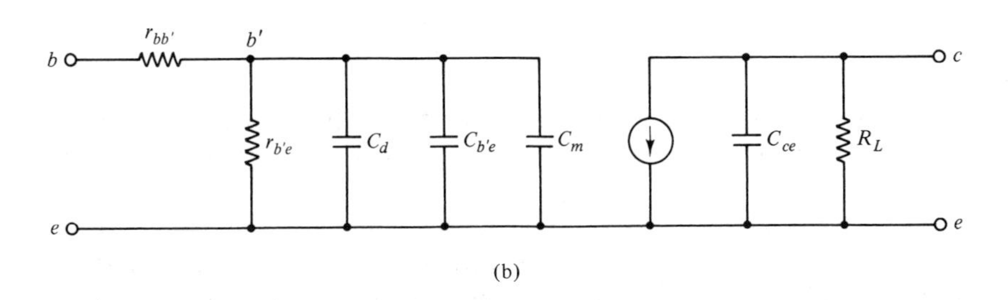

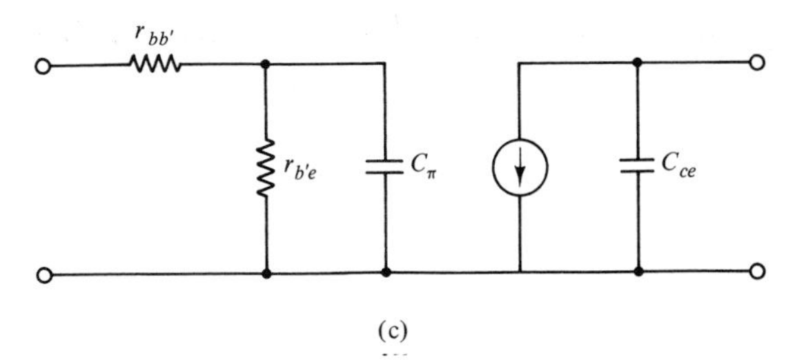

Figure 5-47 High-frequency hybrid-π transistor model. (*a*) Basic high-frequency model, (*b*) high-frequency model with $C_{b'c}$ reflected in base circuit as C_m, and (*c*) simplified model with $C_\pi = C_d + C_{b'e} + C_m$.

Reflecting $C_{b'c}$ into the base circuit, since $Q = CV$, the charge on $C_{b'c}$ is

$$C_{b'c}V_{cb'c} = C_{b'c}V_{b'e}\left(1 + \frac{\beta R_L}{(\beta + 1)r_e}\right)$$

In order to accept the same charge from the base circuit,

$$C_m = C_{b'c}\left(1 + \frac{\beta R_L}{(\beta + 1)r_e}\right)$$

Reflecting $C_{b'c}$ into the collector circuit,

$$V_{C_{b'c}} = V_{ce} - \frac{V_{ce}(\beta + 1)}{\beta R_L}$$

$$V_{C_{b'c}} = V_{ce}\left(1 - \frac{(\beta + 1)r_e}{\beta R_L}\right)$$

Following the logic above,

$$C_{ce} = C_{b'c}\left(1 - \frac{(\beta + 1)r_e}{\beta R_L}\right) \approx C_{b'c}$$

The reactive and parasitic elements in the transistor model of Figure 5-47 must be considered if the transistor is used at frequencies near f_β.

Figure 5-47(*b*) illustrates the model with $C_{b'e}$ reflected into the base circuit as C_m. When the load impedance is being calculated the reflected value of $C_{b'c}$ in the collector circuit C_{ce} must be considered otherwise C_m accounts for the effect of $C_{b'c}$.

Figure 5-48 illustrates the short-circuit current gain on a transistor

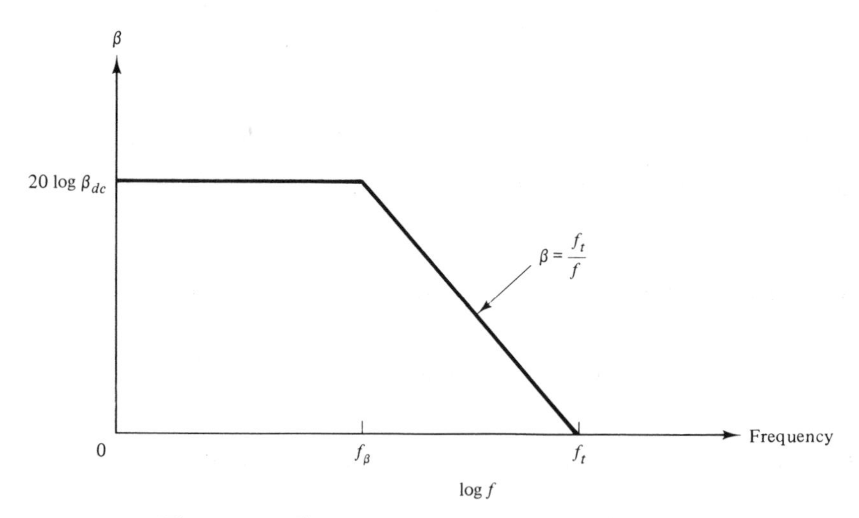

Figure 5-48 Short-circuit current gain versus frequency.

versus frequency. The dc value of β is valid for frequencies up to f_β. At f_β (beta cut-off frequency) the value of beta drops by 3 dB then continues to drop at a rate of 6 db/octave of frequency until $f = f_t$, at which frequency beta = 1 (0 dB). The value f_t is the quantity usually supplied by vendors. Knowing f_t and dc beta, f_β can be easily calculated as

$$f_\beta = \frac{f_t}{\beta}$$

The value of f_β, and therefore which frequencies constitute high frequencies, depends on the transistor geometry, transistor type, and process. A minimum geometry *npn* transistor on a 12-V semicustom process may have an f_β in excess of 5 MHz. A large geometry *npn* transistor on the same process may have an f_β of 2 MHz. A lateral *pnp* transistor on the same process will have an f_β of less than 100 kHz. The f_β of similar devices may be half of these values on a 20-V semicustom process.

Short-circuit current gain is measured using the circuit shown in Figure 5-49. The transistor's base is driven from a high-impedance current source and its collector is shorted to ac ground. It is readily seen from Figure 5-49 that f_β is determined by

$$f_\beta = \frac{1}{2\pi r_{b'e} C_\pi}$$

and since $f_t = \beta f_\beta$

$$f_t = \frac{\beta}{2\pi r_{b'e} C_\pi} \approx \frac{1}{2\pi r_e C_\pi}$$

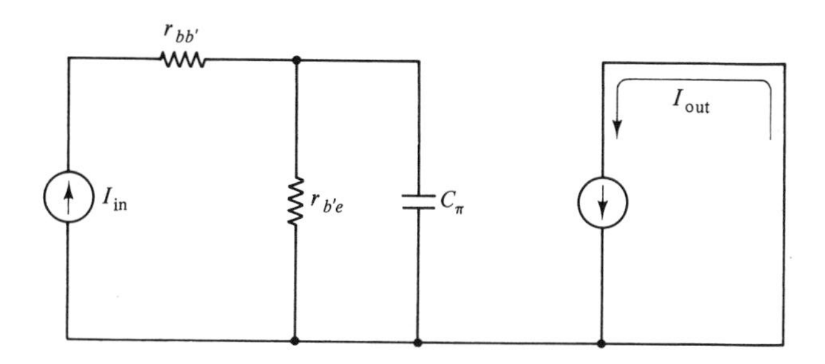

Figure 5-49 Circuit for measuring short-circuit current versus frequency.

Note that since $R_L = 0$, $C_m = C_{b'c}$. Also, $C_{b'e} \ll C_d$ and $C_{b'c} \ll C_d$, so

$$C_\pi \approx C_d$$

High-Frequency Amplifier

Upon close examination of the circuit in Figure 5-49, several things are apparent. With the collector shorted to ground the voltage gain is zero; therefore C_m is equal to $C_{b'c}$ and $C_{b'c}$ reflected into the collector circuit has no effect on frequency response since it is shorted. The base circuit is driven by a high-impedance current source and $r_{bb'}$ has no effect on circuit performance.

Practical circuits, however, will have a finite source impedance and a nonzero resistance in the emitter circuit. The addition of these modifications to the circuit in Figure 5-49 has several effects on the usable frequency response of the circuit. Figure 5-50 illustrates an equivalent circuit of a "practical" common emitter amplifier. A finite source impedance, R_S, added to $r_{bb'}$ appears in parallel with $r_{b'e}$ and C_π. This has the effect of increasing the frequency of the input circuit pole. Adding a resistor in the emitter circuit has the effect of increasing $r_{b'e}$ and reducing the input circuit pole. Adding a collector load resistor increases gain, thereby making C_m larger, which reduces the frequency of the input circuit pole.

In summary, reducing the source impedance has the effect of increasing the circuit's bandwidth. In the limit, $RS = 0$ and the input pole is determined by $r_{bb'}//r_{b'e}$ and C_π. Adding emitter resistance that is not bypassed with a capacitor at the frequency of interest decreases the circuit bandwidth. As R_L is increased, the amplifier gain is increased but C_m is increased by roughly the same amount, thereby decreasing bandwidth.

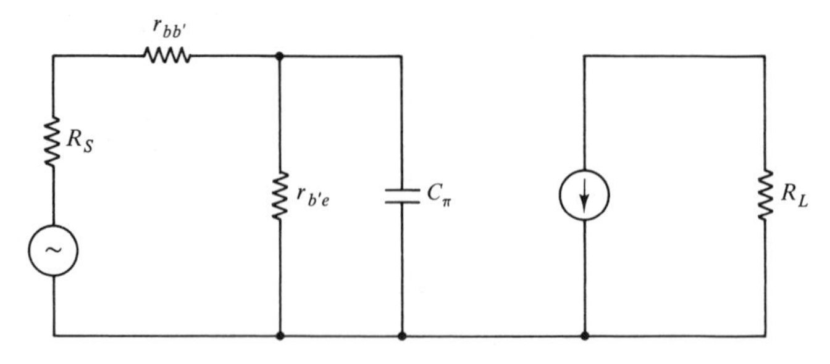

Figure 5-50 Practical amplifier circuit.

Operational Amplifiers

Operational amplifiers derive their name from their use to perform mathematical operations such as integration, differentiation, and summation.

An op amp is a gain block, the functionality of which is modified by feedback. This makes the amplifier easy to adapt to many different uses and makes its characteristics such as closed loop gain and frequency response easy to tailor with very few components.

An ideal op amp, depicted in Figure 5-51(*a*), has zero input offset voltage, an infinite input impedance, zero input bias current, infinite open loop gain, infinite bandwidth, and zero output impedance. Although these ideal characteristics are unachievable, typical op amp characteristics, as shown in Figure 5-51(*b*), come close enough for many applications operating from dc to a few hundred kilohertz.

A typical op amp consists of three major circuit blocks, as shown in Figure 5-52:[14] an input stage, a second gain stage, and an output buffer. This circuit is similar in concept to the circuit of Figure 5-46. If the circuit in Figure 5-46 is modified by using *pnp* transistors as input devices—a current mirror for collector loads on the first stage, a current source load on the second gain stage, and a class B push–pull circuit for the output buffer—the classic op amp circuit shown in Figure 5-53 results. Q_5, Q_6, R_1, and Q_9 form a current source bias string, with R_b setting the bias current. Q_1 and Q_2 are the input differential pair with the current mirror formed by Q_3 and Q_4 serving as collector loads. Q_7 is the second gain stage with current source Q_9 as collector load. C_c is the compensation capacitor providing a dominant pole in the amplifier to ensure stability. Q_8 provides 1 V_{be} of bias between the push–pull output buffers Q_{10} and Q_{11} to limit the crossover distortion to 1 V_{be}.

As an example, we will choose values for R_b, R_1, R_2, and C_c to yield a circuit with a gain bandwidth of 1 MHz, input bias current less than 1 μA, and output drive greater than 5 mA. We will then calculate

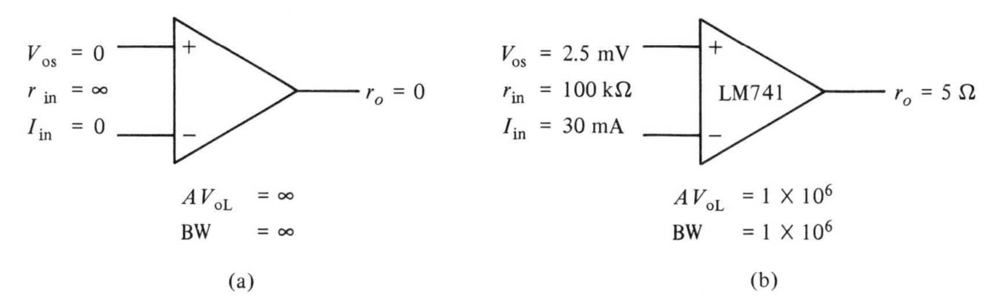

Figure 5-51 Operational amplifier characteristics for (*a*) ideal op amp and (*b*) typical op amp.

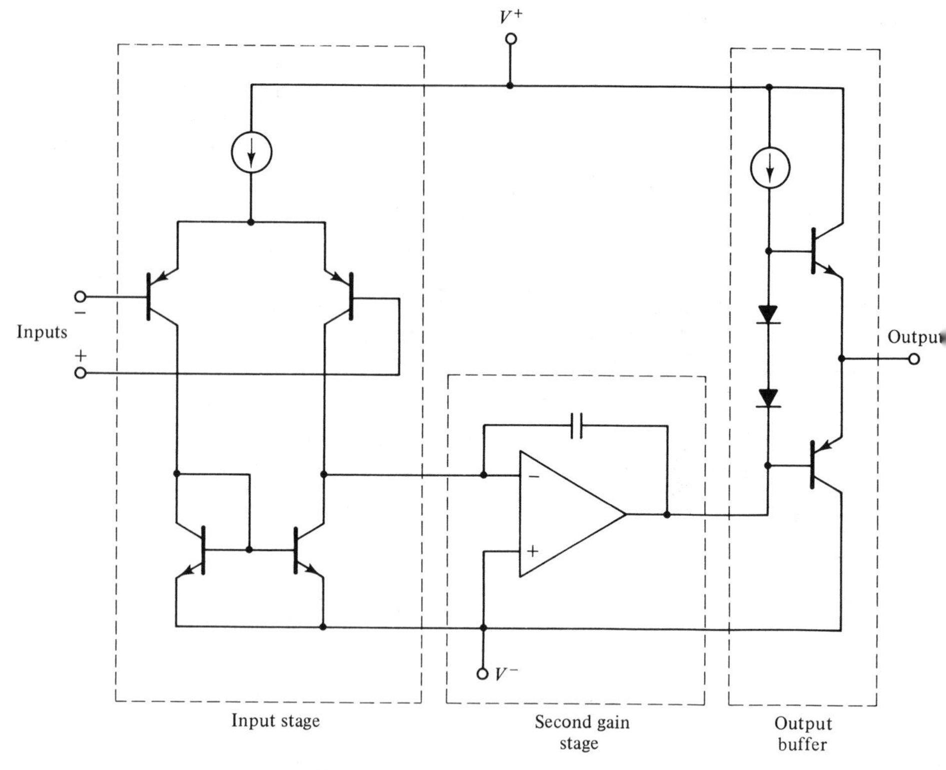

Figure 5-52 Op amp block diagram.

the slew rate and open loop gain for the amplifier. The resulting circuit will be simulated with SPICE to verify the results.

Assumptions:

$$
\begin{aligned}
\beta_{npn} &= 150 \\
\beta_{pnp} &= 70 \text{ (lateral)} \\
\beta_{pnp} &= 90 \text{ (vertical)} \\
Va_{npn} &= 200 \text{ V} \\
Va_{pnp} &= 80 \text{ V} \\
V^{+} &= +5 \text{ V} \\
V^{-} &= -5 \text{ V}
\end{aligned}
$$

Choose Q_5 collector current to be 20 μA. Under balanced conditions, Q_1 and Q_2 will each conduct 10 μA. I_b will be

$$I_b = \frac{10\ \mu\text{A}}{70} = 143 \text{ nA}$$

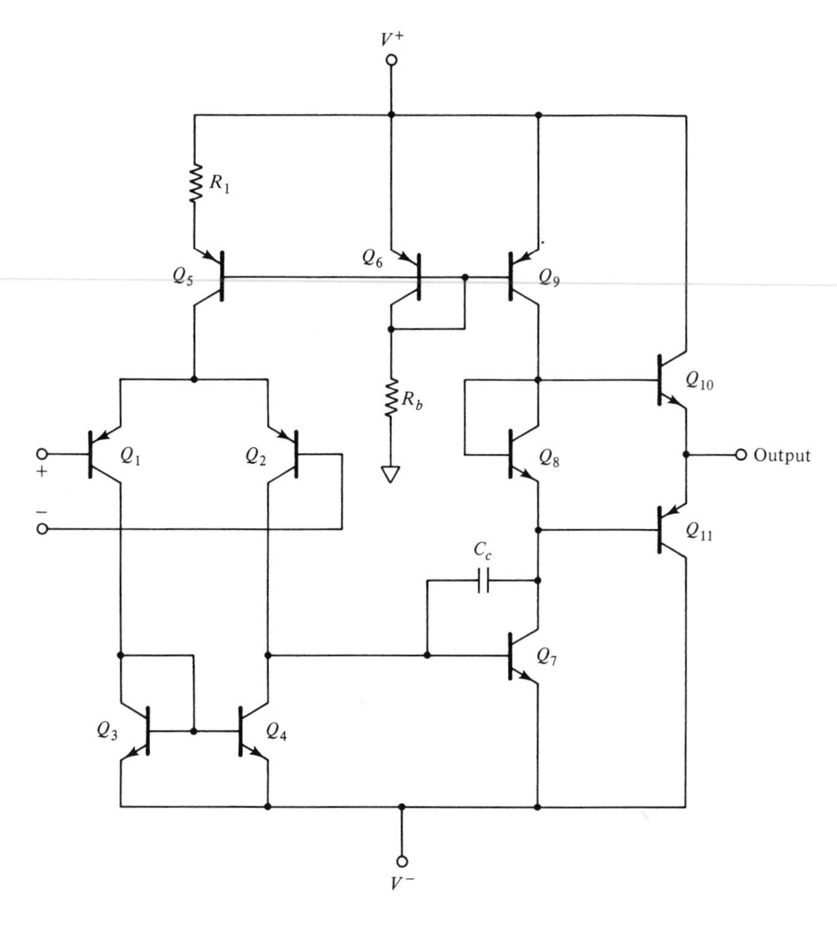

Figure 5-53 Simple op amp.

Q_{11} is a small-geometry vertical *pnp* and its base current will be

$$I_{bQ11} = \frac{5 \text{ mA}}{90} = 55.6 \ \mu\text{A}$$

The base current for Q_{10} under full load conditions will be

$$I_{bQ10} = \frac{5 \text{ mA}}{150} = 33.3 \ \mu\text{A}$$

Let $I_{cQ9} = 50 \ \mu\text{A}$.

Q_7 will have to sink the collector current from Q_9 and the base current from Q_{11}.

$$I_{cQ7} = I_{cQ9} + I_{bQ11} = 50\ \mu A + 55.6\ \mu A = 105.6\ \mu A$$

The base current available to Q_7 must be greater than

$$I_{bQ7} > \frac{105.6\ \mu A}{150} = 0.704\ \mu A$$

Maximum I_bQ_7, available if input stage is hard-switched, is 20 μa, which is more than adequate.

To find the value of R_1 we must calculate the ΔV_{be} between Q_5 and Q_9:

$$\Delta V_{be} = \frac{kT}{q} \ln \frac{I_{cQ_9}}{I_{cQ_5}} = 0.0259 \ln \frac{50\ \mu A}{20\ \mu A} = 23.7\ mV$$

$$R_1 = \frac{23.7\ mV}{20\ \mu A} = 1.2\ k\Omega$$

Calculate the Value for R_b

The total current through R_b is

$$I_{R_b} = I_{cQ_6} + I_{bQ_5} + I_{bQ_6}\ I_{bQ_9}$$

$$I_{R_b} = 50\ \mu A + \frac{20\ \mu A}{70} + \frac{50\ \mu A}{70} + \frac{50\ \mu A}{70} = 51.7\ \mu A$$

$$R_b = \frac{V^+ - V_{beQ_6}}{51.7\ \mu A} = \frac{5\ V - 0.65\ V}{51.7\ \mu A} = 84.1\ k\Omega$$

Now that the bias currents have been determined, we will calculate the open loop gain, C_c, and slew rate response of the amplifier.

Calculate Open Loop Gain

Figure 5-54 is the equivalent circuit to calculate the first-stage gain.

$$V_{in} = I_b(r_{bb'} + r_{b'e})$$

$$V_{out} = \beta I_b R_L$$

$$AV_1 = \frac{V_{out}}{V_{in}} - \frac{\beta R_L}{r_{bb'} + (\beta + 1)r_e} = g_m R_L$$

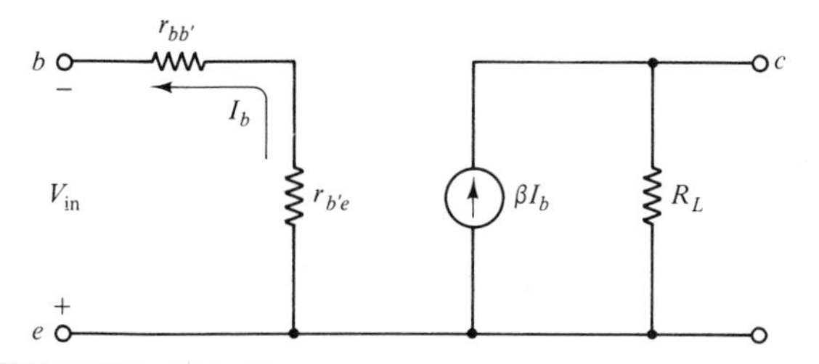

Figure 5-54 First-stage equivalent circuit.

where

$$g_m = \frac{I_{out}}{V_{in}}$$

$$r_e = \frac{kT}{qI_e} \approx \frac{0.0259\ V}{10\ \mu A} = 2.6\ k\Omega$$

Assume $r_{bb'} \approx 200\ \Omega$:

$$AV_1 = \frac{70\ R_L}{200 + (71)(2.6\ K\ \Omega)} \approx \frac{R_L}{r_e}$$

$$R_L = r_{oQ_2}//r_{oQ_4}//r_{inQ_7}$$

$$R_L = \frac{70}{10\ \mu A}//\frac{150}{10\ \mu A}//(151)\frac{kT}{q(50\ \mu A)}$$

$$R_L = 7\ M\Omega//15\ M\Omega//78.2\ k\Omega \approx 78.2\ k\Omega$$

$$AV_1 = \frac{78.2\ k\Omega}{(2.6\ k\Omega)} = 30.1$$

Calculate the Second-Stage Gain

$$V_{in} = I_b(r_{bb'} + r_{b'e})$$

$$V_{out} = \beta I_b R_L$$

$$AV_2 = \frac{\beta R_L}{r_{bb'} + (\beta + 1)r_e} \approx \frac{R_L}{r_e}$$

$$R_L = r_{oQ_9}//r_{oQ_7}//r_{inQ_{11}}$$

Since Q_{11} has a lower beta than Q_{10} and since only one of the devices is on at a time, the worst-case loading will come from the impedance looking into Q_{11}. For this example we will assume a 10-kΩ load on the output.

$$R_L = \frac{80\ \text{V}}{50\ \mu\text{A}} // \frac{200\ \text{V}}{50\ \mu\text{A}} // (91)10\ \text{k}\Omega$$

$$R_L = 1.6\ \text{M}\Omega // 4\ \text{M}\Omega // 910\ \text{k}\Omega \approx 507\ \text{k}\Omega$$

$$AV_2 \approx \frac{R_L}{r_{eQ7}}$$

$$r_{eQ7} = \frac{kT}{qI_e} \approx \frac{0.0259\ \text{V}}{50\ \mu\text{A}} = 518\ \Omega$$

$$AV_2 = \frac{507\ \text{k}\Omega}{(518\ \Omega)} = 978$$

$$AV_{OL} = AV_1 AV_2 = (30.1)(978) = 29.4 \times 10^3$$

$$AV_{OL} = 29.4\ \text{k}\frac{\text{V}}{\text{V}}$$

Calculate the Value of the Compensation Capacitor

Figure 5-55 is the equivalent circuit from which we can calculate the value of the compensation capacitor. The high-frequency current flowing out of the first stage into the base of Q_7 is

$$I = V_{in} g_m$$

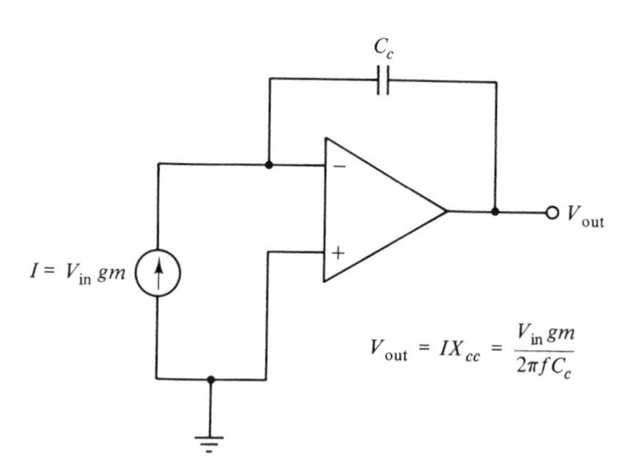

Figure 5-55 Second-stage high-frequency circuit.

From Figure 5-55:

$$V_{out} = IX_{CC} = \frac{V_{in}g_m}{2\pi fC_c}$$

$$AV = \frac{V_{out}}{V_{in}} = \frac{g_m}{2\pi fC_c}$$

It was previously shown that g_m for the first stage is

$$g_m = \frac{1}{r_e}$$

This yields

$$AV = \frac{1}{r_e}\frac{1}{2\pi fC_c}$$

At unity gain $AV = 1$

$$1 = \frac{1}{2\pi r_e fC_c}$$

$$C_c = \frac{1}{2\pi r_e f}$$

$$r_e = \frac{kT}{qI_c} \approx \frac{0.0259\text{ V}}{10\ \mu\text{A}} = 2.6\text{ k}\Omega$$

At $f = 1$ MHz

$$C_c = \frac{1}{2\pi(2.6\text{ k}\Omega)(1\text{ MHz})} = 61.2\text{ pF}$$

Slew Rate

The slew rate of an amplifier is the maximum time rate of change of output voltage. The slew rate is important when the amplifier is required to amplify fast rise time pulses, sinusoids, or any other rapidly moving wave form.

For a sine wave, $V = V_p \sin \omega t$.

The maximum rate of change can be found by taking the derivative and setting it equal to zero and solving for the maximum.

$$\frac{\partial V}{\partial t} = \omega V_p \cos \omega t \text{ (max at } t = 0)$$

$$\text{slew rate} = 2\pi f V_p$$

A sinusoidal waveform has the maximum rate of change as it crosses zero. The ability of an amplifier to operate on a sinusoidal waveform depends on the signal frequency and peak amplitude.

$$I = C\frac{dV}{dt}$$

$$\text{slew rate} = \frac{dV}{dt} = \frac{I}{C_c}$$

where I is the input-stage tail current.

For our example,

$$\text{slew rate} = \frac{I}{C_c} = \frac{20\ \mu\text{A}}{61.2\ \text{pF}} = 0.327\ \frac{\text{V}}{\mu\text{s}}$$

There are anomalies in the slew rate, depending on whether the amplifier is slewing in the positive or negative direction. This depends on the amount of capacitance on the emitters of the input differential pair. For more details see Reference 14.

Some other op amp specifications of importance include input common-mode voltage range and output voltage swing. Figure 5-56 illustrates a common-mode voltage applied to the input of an op amp. The range of this common-mode voltage is limited and depends on the design of the input stage. If the allowable input common-mode voltage range is exceeded, the output of the amplifier can behave unpredictably. The amplifier in Figure 5-53 can have its inputs to within about two diode drops of V^+. When at this potential, the collector-base voltage of Q_5 is zero. The input can be brought to within a V_{be} of V^-. At this potential the input differential pair will have approximately 0 V collector-to-base. If a Wilson or buffered current source was used on the current mirror to improve matching, an additional V_{be} of common-mode range would be lost. On the other hand, if Darlington *pnp* tran-

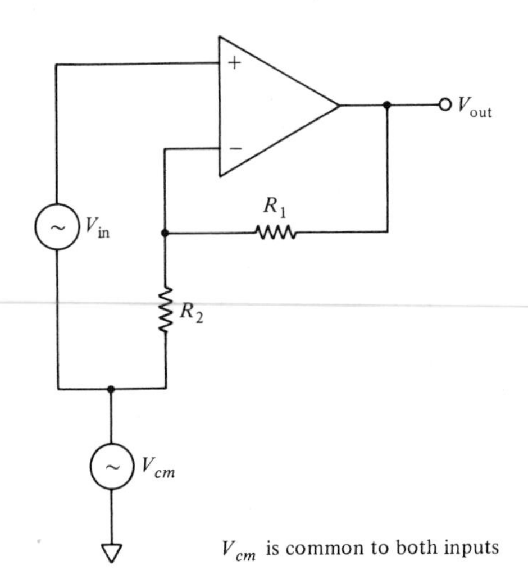

Figure 5-56 Op amp with common-mode voltage.

sistors were used for the input stage, the common-mode voltage range would include the negative supply. Even though a V_{be} of common-mode range was gained on the low end of the range with the use of the Darlingtons, a V_{be} would be lost from the high end of the common-mode range.

The output voltage swing for the schematic in Figure 5-53 is limited to the positive supply voltage minus the saturation voltage of Q_9 and the V_{be} of Q_{10}.

$$V_{out} = V^+ - V_{satQ_9} - V_{beQ_{10}}$$

$$V_{out} \approx V^+ - 0.3\text{ V} - 0.7\text{ V} \approx V^+ - 1\text{ V}$$

The output swing in the negative direction is limited to the negative supply voltage plus the saturation voltage of Q_7 plus the V_{be} of Q_{11}.

$$V_{out} = V^- + V_{satQ_7} + V_{beQ_{11}}$$

$$V_{out} = V^- + 0.3\text{ V} + 0.7\text{ V} \approx V^- + 1\text{ V}$$

If Darlington output devices or buffered current sources are used, the output swing will be reduced.

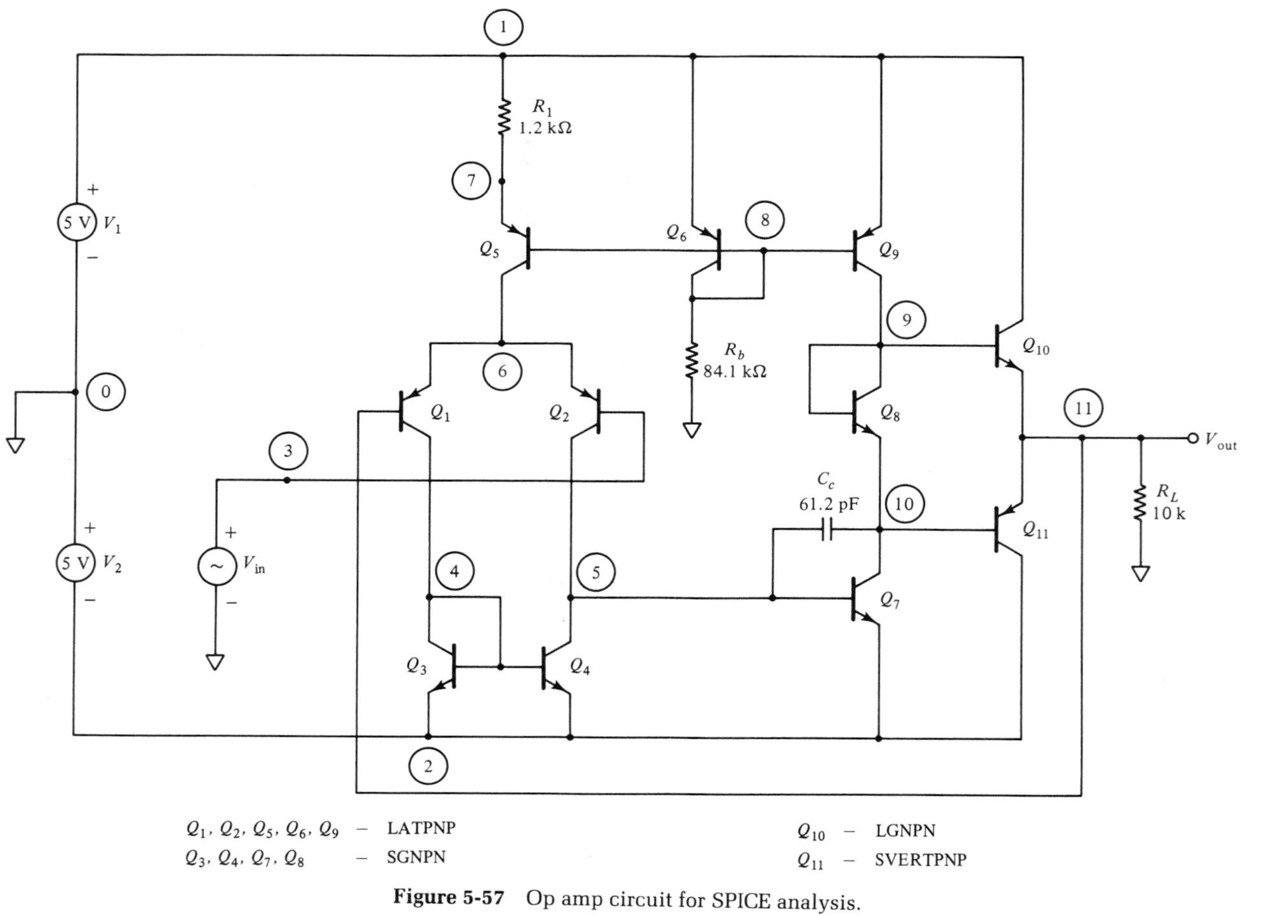

Figure 5-57 Op amp circuit for SPICE analysis.

```
OPERATIONAL AMPLIFIER          BY: PAUL BROWN

***CIRCUIT DESCRIPTION***
*This is a simple pnp input operational amplifier
*connected
*as a unity gain voltage follower.
*
*
***CIRCUIT CONNECTION***

***RESISTORS***

R1   7    1    1.2K
RB   0    8    84.1K
RL   0    11   10K
***CAPACITORS***

CC   5    10   61.2PF
***SEMICONDUCTORS***

Q1   4    11   6    2    LATPNP
Q2   5    3    6    2    LATPNP
Q3   4    4    2    2    SGNPN
Q4   5    4    2    2    SGNPN
Q5   6    8    7    2    LATPNP
Q6   8    8    1    2    LATPNP
Q7   10   5    2    2    SGNPN
Q8   9    9    10   2    SGNPN
Q9   9    8    1    2    LATPNP
Q10  1    9    11   2    LGNPN
Q11  2    10   11   2    SVERTPNP

***SOURCES***

V1   1    0    5V
V2   0    2    5V
VIN 3 0 AC 1 PWL 0  -1 1US -1V 1.001US  1V 16US 1V
+16.001US -1V 31US -1V
***MODEL DEFINITION***
```

Figure 5-58 Simple op amp SPICE input file. (*Figure continues.*)

We will measure the performance of the circuit in Figure 5-53 by performing an ac and transient SPICE analysis. The circuit with node numbers indicated is illustrated in Figure 5-57. The SPICE input file is listed in Figure 5-58 and the resulting output file in Figure 5-59.

Figure 5-60 illustrates the output transient response of the circuit

```
.MODEL    SGNPN     NPN(IS=5E-16 BF=150 VAF=200
+IKF=1E-3 ISE=1E-13 ISC=1E-12 RB=150 RE=2 RC=50
+CJE=2.5E-12 VJE=0.6 TF=0.33E-9 CJC=2E-12 TR=10E-9
+CJS=2E-12 XTB=2.0)

.MODEL    LGNPN     NPN(IS=5E-16 BF=150 VAF=150
+IKF=10E-3 ISE=1E-13 ISC=3E-12 RB=5.0 RE=1.0 RC=10
+CJE=12E-12 VJE=0.6 TF=0.5E-9 CJC=10E-12 TR=15E-9
+CJS=10E-12 XTB=2.0)

.MODEL    LATPNP    PNP(BF=70 VAF=80 IKF=50E-6
+ISE=1E-13 ISC=3.0E-12 RB=300 RE=10.0 RC=200
+CJE=1.0E-12 VJE=0.65 TF=33.0E-9 CJC=2E-12 TR=75E-9
+CJS=2E-12 XTB=2.0)

.MODEL    SVERTPNP PNP(IS=1.0E-15 BF=90 VAF=80
+IKF=1E-3 ISE=1E-13 ISC=3.0E-12 RB=300 RE=10 RC=100
+CJE=1.0E-12 VJE=0.65 TF=20.0E-9 TR=60E-9 XTB=2.0)


***ANALYSIS CONTROL***
.AC DEC 10 10 10MEG
.TRAN  250NS  31US

***OUTPUT CONTROL***
.PRINT AC VDB(11) VP(11)
.PRINT TRAN V(11)  V(3)

.END
```

Figure 5-58 *(Continued)*

in Figure 5-57. Notice that the output rate of change is not instantaneous but limited by the compensation capacitor and the g_m of the input stage. Also notice the crossover distortion at zero volts. This is due to the class B output stage. There is a $1 V_{be}$ difference between the point at which Q_{10} turns off and Q_{11} turns on and vice versa.

Figure 5-61 illustrates the calculation of slew rate. The predicted slew rate was 0.327 V/μs. The simulated value is

$$\text{slew rate} = \frac{0.752 \text{ V}}{2.45\ \mu\text{s}} = 0.307\ \frac{\text{V}}{\mu\text{s}}$$

The difference is due to capacitive elements in the computer model that were not considered in the hand calculation. Given this value of slew

```
******* 2/ 3/91 ******* IS SPICE v1.5M 4/14/89 *******10:16:32*****

OPERATIONAL AMPLIFIER           BY: PAUL BROWN

****     INPUT LISTING                    TEMPERATURE =   27.000 DEG C

***********************************************************************



***CIRCUIT DESCRIPTION***
*This is a simple pnp input operational amplifier *connected
*as a unity gain voltage follower.
*
*
***CIRCUIT CONNECTION***

***RESISTORS***

R1      7       1       1.2K
RB      0       8       84.1K
RL      0       11      10K
***CAPACITORS***

CC      5       10      61.2PF
***SEMICONDUCTORS***

Q1      4       11      6       2       LATPNP
Q2      5       3       6       2       LATPNP
Q3      4       4       2       2       SGNPN
Q4      5       4       2       2       SGNPN
Q5      6       8       7       2       LATPNP
Q6      8       8       1       2       LATPNP
Q7      10      5       2       2       SGNPN
Q8      9       9       10      2       SGNPN
Q9      9       8       1       2       LATPNP
Q10     1       9       11      2       LGNPN
Q11     2       10      11      2       SVERTPNP

***SOURCES***

V1      1       0       5V
V2      0       2       5V
VIN 3 0 AC 1 PWL 0  -1 1US -1V 1.001US  1V 16US 1V +16.001US -1V 31US -1V
***MODEL DEFINITION***

.MODEL  SGNPN   NPN(IS=5E-16 BF=150 VAF=200 +IKF=1E-3 ISE=1E-13 ISC=1E-12 RB=150

.MODEL  LGNPN   NPN(IS=5E-16 BF=150 VAF=150 +IKF=10E-3 ISE=1E-13 ISC=3E-12 RB=5.

.MODEL  LATPNP  PNP(BF=70 VAF=80 IKF=50E-6 +ISE=1E-13 ISC=3.0E-12 RB=300 RE=10.0

.MODEL  SVERTPNP        PNP(IS=1.0E-15 BF=90 VAF=80 +IKF=1E-3 ISE=1E-13 ISC=3.0E


***ANALYSIS CONTROL***
.AC DEC 10 10 10MEG
.TRAN  250NS  31US

***OUTPUT CONTROL***
.PRINT AC VDB(11) VP(11)
```

Figure 5-59 Simple op amp SPICE output file. (*Figure continues.*)

```
.PRINT TRAN V(11)  V(3)

.END
******* 2/ 3/91 ******* IS SPICE v1.5M 4/14/89 *******10:16:32*****

OPERATIONAL AMPLIFIER           BY: PAUL BROWN

****      BJT MODEL PARAMETERS              TEMPERATURE =   27.000 DEG C

************************************************************************



                SGNPN      LGNPN      LATPNP     SVERTPNP
TYPE            NPN        NPN        PNP        PNP
IS          5.00D-16   5.00D-16   1.00D-16   1.00D-15
BF           150.000    150.000     70.000     90.000
NF             1.000      1.000      1.000      1.000
VAF         2.00D+02   1.50D+02   8.00D+01   8.00D+01
IKF         0.00D-01   0.00D-01   5.00D-05   0.00D-01
ISE         1.00D-13   1.00D-13   0.00D-01   1.00D-13
BR             1.000      1.000      1.000      1.000
NR             1.000      1.000      1.000      1.000
ISC         1.00D-12   3.00D-12   3.00D-12   3.00D-15
C4              .000      0.000       .000      3.000
RB           150.000      5.000    300.000       .000
RE              .000       .000     10.000       .000
******* 2/ 3/91 ******* IS SPICE v1.5M 4/14/89 *******10:16:32*****

OPERATIONAL AMPLIFIER           BY: PAUL BROWN

****      SMALL SIGNAL BIAS SOLUTION        TEMPERATURE =   27.000 DEG C

************************************************************************


 NODE   VOLTAGE      NODE   VOLTAGE      NODE   VOLTAGE      NODE   VOLTAGE


(  1)    5.0000     (  2)   -5.0000     (  3)   -1.0000     (  4)   -4.3881

(  5)   -4.3428     (  6)     -.3357    (  7)    4.9715     (  8)    4.2862

(  9)     -.9956    ( 10)   -1.6503     ( 11)     -.9971



    VOLTAGE SOURCE CURRENTS

    NAME         CURRENT


    V1        -1.252D-04

    V2        -1.738D-04

    VIN        2.023D-07

    TOTAL POWER DISSIPATION   1.50D-03  WATTS
```

Figure 5-59 *(Continued)*

```
******* 2/ 3/91 ******* IS SPICE v1.5M 4/14/89 *******10:16:32*****

OPERATIONAL AMPLIFIER               BY: PAUL BROWN

****      OPERATING POINT INFORMATION       TEMPERATURE =   27.000 DEG C

***********************************************************************



**** BIPOLAR JUNCTION TRANSISTORS

             Q1         Q2         Q3         Q4         Q5         Q6         Q7
MODEL      LATPNP     LATPNP     SGNPN      SGNPN      LATPNP     LATPNP     SGNPN
IB        -1.81E-07 -2.02E-07  7.69E-07   7.69E-07  -4.54E-07 -1.34E-06  2.60E-06
IC        -1.09E-05 -1.20E-05  9.39E-06   9.39E-06  -2.33E-05 -4.79E-05  5.41E-05
VBE          -.661      -.664       .612       .612      -.685      -.714       .657
VBC          3.391      3.343       .000      -.045      4.622       .000     -2.693
VCE         -4.052     -4.007       .612       .657     -5.307      -.714      3.350
BETADC      60.320     59.285     12.209     12.212     51.391     35.765     20.752
GM         3.60E-04   3.90E-04   3.63E-04   3.63E-04   6.90E-04   1.24E-03   2.09E-03
RPI        1.43E+05   1.28E+05   4.85E+04   4.85E+04   5.70E+04   1.93E+04   1.39E+04
RX         3.00E+02   3.00E+02   1.50E+02   1.50E+02   3.00E+02   3.00E+02   1.50E+02
RO         7.63E+06   6.95E+06   2.13E+07   2.13E+07   3.63E+06   1.67E+06   3.75E+06
CPI        0.00E-01   0.00E-01   0.00E-01   0.00E-01   0.00E-01   0.00E-01   0.00E-01
CMU        0.00E-01   0.00E-01   0.00E-01   0.00E-01   0.00E-01   0.00E-01   0.00E-01
CBX        0.00E-01   0.00E-01   0.00E-01   0.00E-01   0.00E-01   0.00E-01   0.00E-01
CCS        0.00E-01   0.00E-01   0.00E-01   0.00E-01   0.00E-01   0.00E-01   0.00E-01
BETAAC      51.388     49.923     17.595     17.599     39.335     24.006     29.135
FT         5.73E+15   6.21E+15   5.78E+15   5.78E+15   1.10E+16   1.98E+16   3.33E+16


             Q8         Q9         Q10        Q11
MODEL      SGNPN      LATPNP     LGNPN      SVERTPNP
IB         2.44E-06 -1.34E-06 -8.99E-12 -3.08E-06
IC         4.85E-05 -5.10E-05  1.52E-11 -9.68E-05
VBE           .655      -.714       .001      -.653
VBC           .000      5.282     -5.996      3.350
VCE           .655     -5.996      5.997     -4.003
BETADC      19.925     38.149     -1.694     31.394
GM         1.88E-03   1.32E-03 -1.88E-14   3.74E-03
RPI        1.49E+04   1.94E+04   3.72E+11   1.08E+04
RX         1.50E+02   3.00E+02   5.00E+00   0.00E-01
RO         4.12E+06   1.67E+06   9.26E+11   8.61E+05
CPI        0.00E-01   0.00E-01   0.00E-01   0.00E-01
CMU        0.00E-01   0.00E-01   0.00E-01   0.00E-01
CBX        0.00E-01   0.00E-01   0.00E-01   0.00E-01
CCS        0.00E-01   0.00E-01   0.00E-01   0.00E-01
BETAAC      28.023     25.613      -.007     40.326
FT         2.99E+16   2.11E+16 -2.99E+05   5.96E+16
******* 2/ 3/91 ******* IS SPICE v1.5M 4/14/89 *******10:16:32*****

OPERATIONAL AMPLIFIER               BY: PAUL BROWN

****      AC ANALYSIS                       TEMPERATURE =   27.000 DEG C

***********************************************************************



     FREQ        VDB(11)     VP(11)
```

Figure 5-59 (*Continued*)

```
1.00000E+01    -2.540E-03  -6.631E-04
1.25893E+01    -2.540E-03  -8.347E-04
1.58489E+01    -2.540E-03  -1.051E-03
1.99526E+01    -2.540E-03  -1.323E-03
2.51189E+01    -2.540E-03  -1.666E-03
3.16228E+01    -2.540E-03  -2.097E-03
3.98107E+01    -2.540E-03  -2.640E-03
5.01187E+01    -2.540E-03  -3.323E-03
6.30957E+01    -2.540E-03  -4.184E-03
7.94328E+01    -2.540E-03  -5.267E-03
1.00000E+02    -2.540E-03  -6.631E-03
1.25893E+02    -2.540E-03  -8.347E-03
1.58489E+02    -2.540E-03  -1.051E-02
1.99526E+02    -2.541E-03  -1.323E-02
2.51189E+02    -2.540E-03  -1.666E-02
3.16228E+02    -2.541E-03  -2.097E-02
3.98107E+02    -2.541E-03  -2.640E-02
5.01187E+02    -2.541E-03  -3.323E-02
6.30957E+02    -2.542E-03  -4.184E-02
7.94328E+02    -2.543E-03  -5.267E-02
1.00000E+03    -2.544E-03  -6.631E-02
1.25893E+03    -2.547E-03  -8.347E-02
1.58489E+03    -2.550E-03  -1.051E-01
1.99526E+03    -2.556E-03  -1.323E-01
2.51189E+03    -2.565E-03  -1.666E-01
3.16228E+03    -2.580E-03  -2.097E-01
3.98107E+03    -2.603E-03  -2.640E-01
5.01187E+03    -2.640E-03  -3.323E-01
6.30957E+03    -2.698E-03  -4.184E-01
7.94328E+03    -2.790E-03  -5.267E-01
1.00000E+04    -2.936E-03  -6.630E-01
1.25893E+04    -3.168E-03  -8.347E-01
1.58489E+04    -3.535E-03  -1.051E+00
1.99526E+04    -4.116E-03  -1.323E+00
2.51189E+04    -5.037E-03  -1.665E+00
3.16228E+04    -6.496E-03  -2.096E+00
3.98107E+04    -8.808E-03  -2.639E+00
5.01187E+04    -1.247E-02  -3.321E+00
6.30957E+04    -1.827E-02  -4.179E+00
7.94328E+04    -2.744E-02  -5.258E+00
1.00000E+05    -4.193E-02  -6.613E+00
1.25893E+05    -6.478E-02  -8.312E+00
1.58489E+05    -1.008E-01  -1.044E+01
1.99526E+05    -1.571E-01  -1.309E+01
2.51189E+05    -2.448E-01  -1.638E+01
3.16228E+05    -3.801E-01  -2.044E+01
3.98107E+05    -5.855E-01  -2.537E+01
5.01187E+05    -8.908E-01  -3.128E+01
6.30957E+05    -1.331E+00  -3.818E+01
7.94328E+05    -1.942E+00  -4.604E+01
1.00000E+06    -2.747E+00  -5.467E+01
1.25893E+06    -3.750E+00  -6.386E+01
1.58489E+06    -4.927E+00  -7.335E+01
1.99526E+06    -6.228E+00  -8.298E+01
2.51189E+06    -7.584E+00  -9.265E+01
3.16228E+06    -8.919E+00  -1.023E+02
3.98107E+06    -1.016E+01  -1.119E+02
5.01187E+06    -1.125E+01  -1.212E+02
6.30957E+06    -1.214E+01  -1.301E+02
7.94328E+06    -1.284E+01  -1.384E+02
```

Figure 5-59 *(Continued)*

```
 1.00000E+07     -1.335E+01  -1.457E+02

******* 2/ 3/91 ******* IS SPICE v1.5M 4/14/89 *******10:16:32*****

OPERATIONAL AMPLIFIER             BY: PAUL BROWN

****      INITIAL TRANSIENT SOLUTION       TEMPERATURE =   27.000 DEG C

***********************************************************************


 NODE   VOLTAGE      NODE   VOLTAGE      NODE   VOLTAGE      NODE   VOLTAGE


 (  1)     5.0000    (  2)    -5.0000    (  3)    -1.0000    (  4)    -4.3881

 (  5)    -4.3428    (  6)     -.3357    (  7)     4.9715    (  8)     4.2862

 (  9)     -.9956    ( 10)    -1.6503    ( 11)     -.9971



    VOLTAGE SOURCE CURRENTS

    NAME         CURRENT


    V1         -1.252D-04

    V2         -1.738D-04

    VIN         2.023D-07


    TOTAL POWER DISSIPATION   1.50D-03  WATTS
******* 2/ 3/91 ******* IS SPICE v1.5M 4/14/89 *******10:16:32*****

OPERATIONAL AMPLIFIER             BY: PAUL BROWN

****      OPERATING POINT INFORMATION      TEMPERATURE =   27.000 DEG C

***********************************************************************



**** BIPOLAR JUNCTION TRANSISTORS


             Q1         Q2         Q3         Q4         Q5         Q6         Q7
MODEL      LATPNP     LATPNP     SGNPN      SGNPN      LATPNP     LATPNP     SGNPN
IB        -1.81E-07 -2.02E-07  7.69E-07   7.69E-07 -4.54E-07 -1.34E-06  2.60E-06
IC        -1.09E-05 -1.20E-05  9.39E-06   9.39E-06 -2.33E-05 -4.79E-05  5.41E-05
VBE           -.661      -.664       .612       .612      -.685      -.714       .657
VBC           3.391      3.343       .000      -.045      4.622       .000     -2.693
VCE          -4.052     -4.007       .612       .657     -5.307      -.714      3.350
BETADC       60.320     59.285     12.209     12.212     51.391     35.765     20.752


             Q8         Q9         Q10        Q11
MODEL      SGNPN      LATPNP     LGNPN      SVERTPNP
IB         2.44E-06 -1.34E-06 -8.99E-12 -3.08E-06
```

Figure 5-59 *(Continued)*

```
IC          4.85E-05 -5.10E-05  1.52E-11 -9.68E-05
VBE             .655      -.714      .001     -.653
VBC             .000      5.282    -5.996     3.350
VCE             .655     -5.996     5.997    -4.003
BETADC        19.925     38.149    -1.694    31.394
******* 2/ 3/91 ******* IS SPICE v1.5M 4/14/89 *******10:16:32*****

OPERATIONAL AMPLIFIER              BY: PAUL BROWN

****      TRANSIENT ANALYSIS                  TEMPERATURE =   27.000 DEG C

***********************************************************************


    TIME         V(11)         V(3)


 0.00000E-01     -9.971E-01  -1.000E+00
 2.50000E-07     -9.971E-01  -1.000E+00
 5.00000E-07     -9.971E-01  -1.000E+00
 7.50000E-07     -9.971E-01  -1.000E+00
 1.00000E-06     -9.971E-01  -1.000E+00
 1.25000E-06     -9.230E-01   1.000E+00
 1.50000E-06     -8.360E-01   1.000E+00
 1.75000E-06     -7.495E-01   1.000E+00
 2.00000E-06     -6.631E-01   1.000E+00
 2.25000E-06     -5.779E-01   1.000E+00
 2.50000E-06     -4.927E-01   1.000E+00
 2.75000E-06     -4.088E-01   1.000E+00
 3.00000E-06     -3.261E-01   1.000E+00
 3.25000E-06     -2.439E-01   1.000E+00
 3.50000E-06     -1.691E-01   1.000E+00
 3.75000E-06     -9.422E-02   1.000E+00
 4.00000E-06     -4.751E-02   1.000E+00
 4.25000E-06     -2.014E-02   1.000E+00
 4.50000E-06      4.173E-03   1.000E+00
 4.75000E-06      4.367E-03   1.000E+00
 5.00000E-06      4.561E-03   1.000E+00
 5.25000E-06      2.079E-02   1.000E+00
 5.50000E-06      4.634E-02   1.000E+00
 5.75000E-06      7.919E-02   1.000E+00
 6.00000E-06      1.527E-01   1.000E+00
 6.25000E-06      2.261E-01   1.000E+00
 6.50000E-06      3.049E-01   1.000E+00
 6.75000E-06      3.862E-01   1.000E+00
 7.00000E-06      4.680E-01   1.000E+00
 7.25000E-06      5.514E-01   1.000E+00
 7.50000E-06      6.348E-01   1.000E+00
 7.75000E-06      7.185E-01   1.000E+00
 8.00000E-06      8.023E-01   1.000E+00
 8.25000E-06      8.787E-01   1.000E+00
 8.50000E-06      9.306E-01   1.000E+00
 8.75000E-06      9.825E-01   1.000E+00
 9.00000E-06      9.981E-01   1.000E+00
 9.25000E-06      1.002E+00   1.000E+00
 9.50000E-06      1.004E+00   1.000E+00
 9.75000E-06      1.003E+00   1.000E+00
 1.00000E-05      1.001E+00   1.000E+00
 1.02500E-05      1.002E+00   1.000E+00
 1.05000E-05      1.002E+00   1.000E+00
 1.07500E-05      1.002E+00   1.000E+00
 1.10000E-05      1.002E+00   1.000E+00
 1.12500E-05      1.002E+00   1.000E+00
 1.15000E-05      1.002E+00   1.000E+00
 1.17500E-05      1.002E+00   1.000E+00
 1.20000E-05      1.002E+00   1.000E+00
 1.22500E-05      1.002E+00   1.000E+00
 1.25000E-05      1.002E+00   1.000E+00
 1.27500E-05      1.002E+00   1.000E+00
 1.30000E-05      1.002E+00   1.000E+00
 1.32500E-05      1.002E+00   1.000E+00
 1.35000E-05      1.002E+00   1.000E+00
```

Figure 5-59 *(Continued)*

```
1.37500E-05     1.002E+00   1.000E+00
1.40000E-05     1.002E+00   1.000E+00
1.42500E-05     1.002E+00   1.000E+00
1.45000E-05     1.002E+00   1.000E+00
1.47500E-05     1.002E+00   1.000E+00
1.50000E-05     1.002E+00   1.000E+00
1.52500E-05     1.002E+00   1.000E+00
1.55000E-05     1.002E+00   1.000E+00
1.57500E-05     1.002E+00   1.000E+00
1.60000E-05     1.002E+00   1.000E+00
1.62500E-05     9.324E-01  -1.000E+00
1.65000E-05     8.538E-01  -1.000E+00
1.67500E-05     7.756E-01  -1.000E+00
1.70000E-05     6.974E-01  -1.000E+00
1.72500E-05     6.201E-01  -1.000E+00
1.75000E-05     5.427E-01  -1.000E+00
1.77500E-05     4.663E-01  -1.000E+00
1.80000E-05     3.905E-01  -1.000E+00
1.82500E-05     3.151E-01  -1.000E+00
1.85000E-05     2.432E-01  -1.000E+00
1.87500E-05     1.714E-01  -1.000E+00
1.90000E-05     1.108E-01  -1.000E+00
1.92500E-05     5.797E-02  -1.000E+00
1.95000E-05     1.053E-02  -1.000E+00
1.97500E-05     6.090E-03  -1.000E+00
2.00000E-05     1.652E-03  -1.000E+00
2.02500E-05    -5.046E-04  -1.000E+00
2.05000E-05    -1.328E-03  -1.000E+00
2.07500E-05    -7.887E-03  -1.000E+00
2.10000E-05    -4.664E-02  -1.000E+00
2.12500E-05    -8.539E-02  -1.000E+00
2.15000E-05    -1.448E-01  -1.000E+00
2.17500E-05    -2.144E-01  -1.000E+00
2.20000E-05    -2.850E-01  -1.000E+00
2.22500E-05    -3.597E-01  -1.000E+00
2.25000E-05    -4.344E-01  -1.000E+00
2.27500E-05    -5.104E-01  -1.000E+00
2.30000E-05    -5.869E-01  -1.000E+00
2.32500E-05    -6.635E-01  -1.000E+00
2.35000E-05    -7.406E-01  -1.000E+00
2.37500E-05    -8.178E-01  -1.000E+00
2.40000E-05    -8.785E-01  -1.000E+00
2.42500E-05    -9.338E-01  -1.000E+00
2.45000E-05    -9.773E-01  -1.000E+00
2.47500E-05    -9.893E-01  -1.000E+00
2.50000E-05    -1.001E+00  -1.000E+00
2.52500E-05    -1.001E+00  -1.000E+00
2.55000E-05    -9.973E-01  -1.000E+00
2.57500E-05    -9.952E-01  -1.000E+00
2.60000E-05    -9.965E-01  -1.000E+00
2.62500E-05    -9.977E-01  -1.000E+00
2.65000E-05    -9.975E-01  -1.000E+00
2.67500E-05    -9.971E-01  -1.000E+00
2.70000E-05    -9.969E-01  -1.000E+00
2.72500E-05    -9.970E-01  -1.000E+00
2.75000E-05    -9.972E-01  -1.000E+00
2.77500E-05    -9.972E-01  -1.000E+00
2.80000E-05    -9.971E-01  -1.000E+00
2.82500E-05    -9.971E-01  -1.000E+00
2.85000E-05    -9.971E-01  -1.000E+00
2.87500E-05    -9.971E-01  -1.000E+00
2.90000E-05    -9.971E-01  -1.000E+00
2.92500E-05    -9.971E-01  -1.000E+00
2.95000E-05    -9.971E-01  -1.000E+00
2.97500E-05    -9.971E-01  -1.000E+00
3.00000E-05    -9.971E-01  -1.000E+00
3.02500E-05    -9.971E-01  -1.000E+00
3.05000E-05    -9.971E-01  -1.000E+00
3.07500E-05    -9.971E-01  -1.000E+00
3.10000E-05    -9.971E-01  -1.000E+00


          JOB CONCLUDED
           TOTAL JOB TIME            50.22
```

Figure 5-59 (*Continued*)

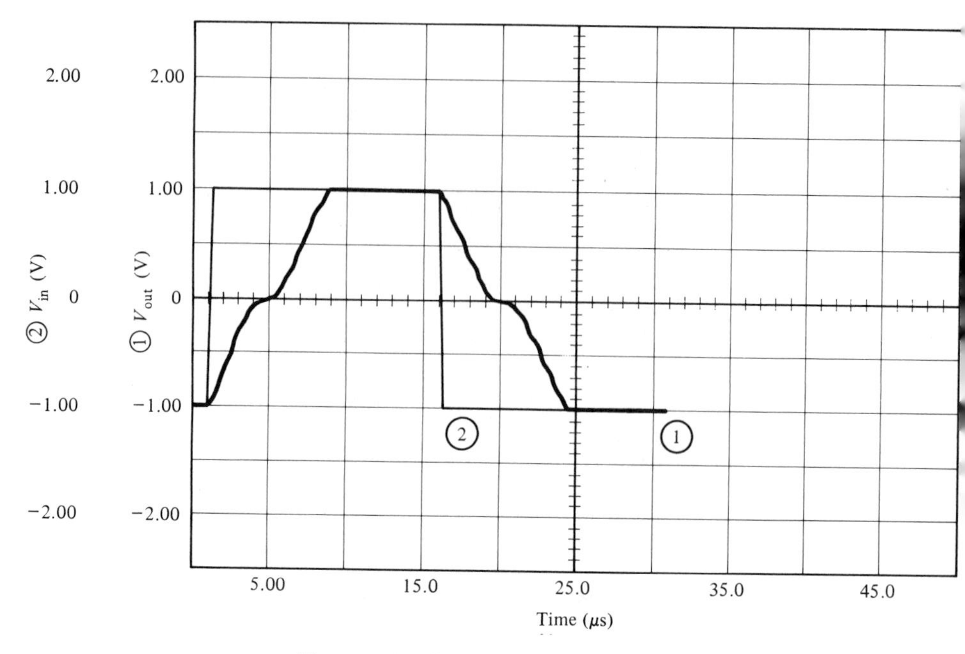

Figure 5-60 Output transient response.

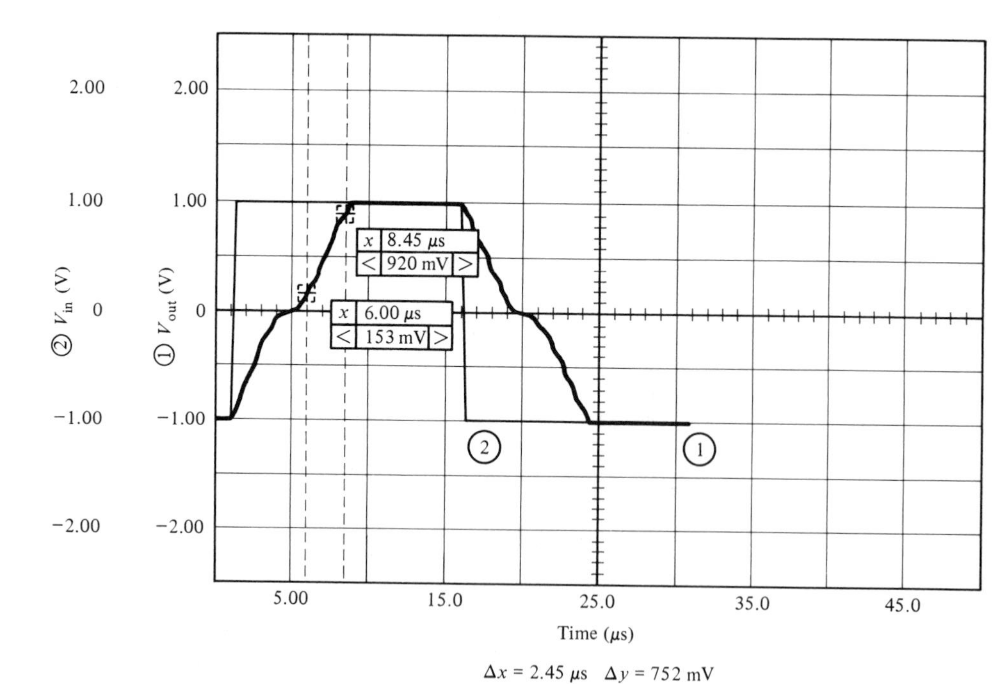

Figure 5-61 Slew-rate calculation.

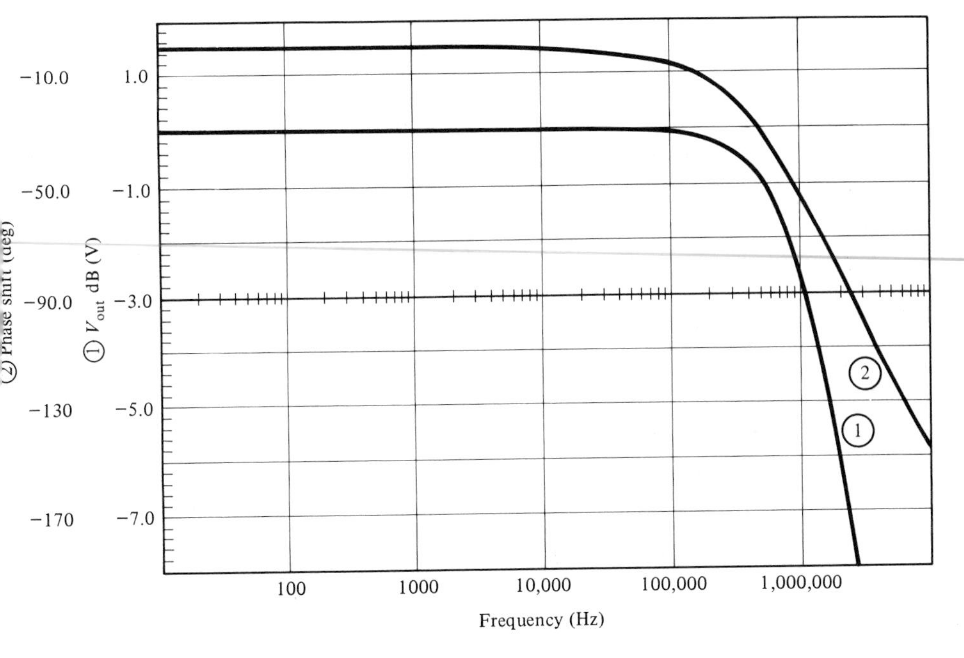

Figure 5-62 Bode plot.

rate, the $1 V_{be}$ crossover distortion should last approximately 2 μs. This agrees with the output curve.

Figure 5-62 is the Bode plot for the simple op amp connected in the unity gain follower configuration and driving a 10 kΩ load (Figure 5-57). The −3 dB frequency is 1 MHz, as we calculated earlier.

There are many different op amp configurations possible. Higher performance generally means higher complexity. When designing standard product op amps, higher complexity is warranted since the amplifier must be designed to operate under the widest possible range of conditions. As part of an ASIC design, the application of the circuit will generally be fixed and well-known. This can allow a simpler design.

Figure 5-63 shows some simple ways the amplifier in Figure 5-53 can be improved. The use of buffer transistor Q_3 reduces base current error in the current mirror. R_1 and R_2 provide a means for offset adjustment. Adding Q_6 improves the input impedance to the second stage. Using R_5, R_6, and Q_8 as a V_{be} multiplier in place of a single diode removes the 1 V_{be} crossover distortion in the output stage. If the values of R_5 and R_6 are correctly chosen, Q_9 and Q_{10} will conduct a small amount of current in the quiescent state. The output stage then becomes class

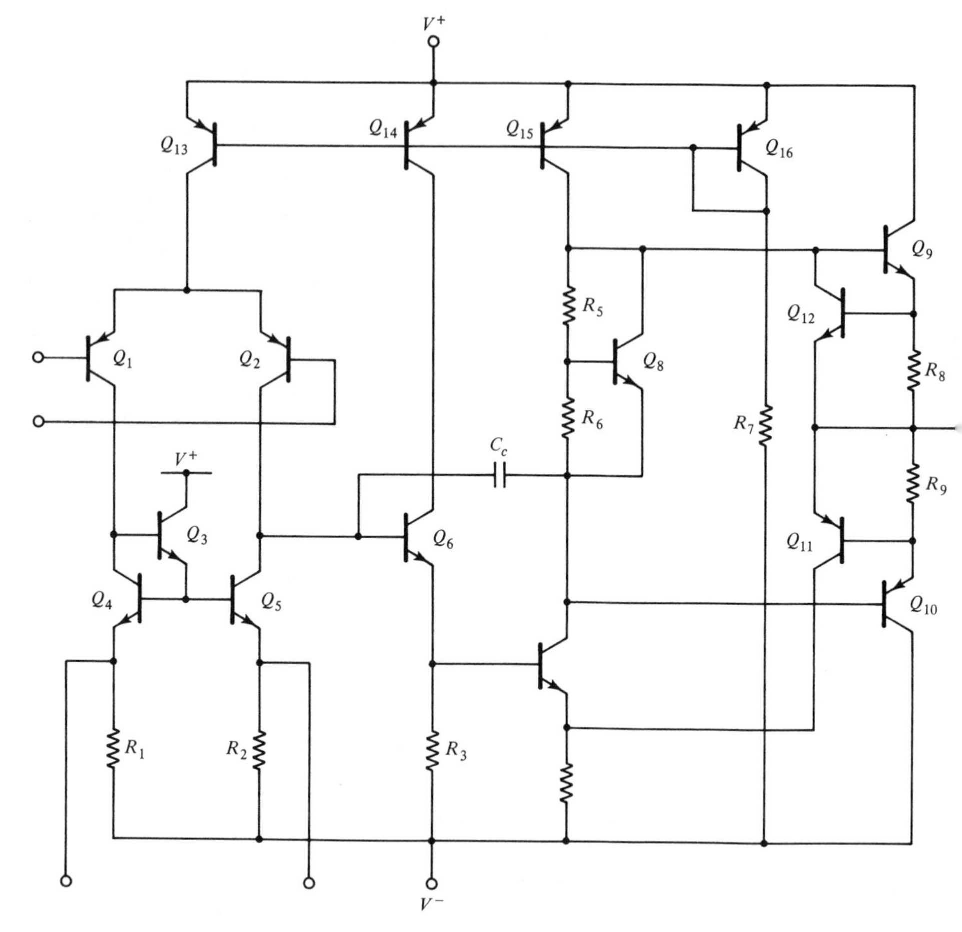

Figure 5-63 An improved op amp.

AB. R_8 and Q_{12} provide current limit protection for Q_9 and R_9, R_4, and Q_{11} provide current limit protection for Q_{10}. This amplifier has some nice features but is significantly more complex than the amplifier in Figure 5-53.

Reference 14 is a classic work on operational amplifiers. Additional insight can be obtained from References 15 and 16.

Simulating a large system-level circuit made up of many op amps can become difficult. This is especially true if more complex op amps, like the circuit illustrated in Figure 5-63, are used. The sheer number of components makes the simulation unwieldy. The simulation time increases and convergence problems become more frequent. The files become so large, in some cases, that small computers are no longer adequate.

The concept of macromodels greatly alleviates this problem. The circuit for an op amp macromodel is shown in Figure 5-64.[17] The model elements that emulate the various circuit functions are indicated on the Figure. Macromodels simulate the functional behavior of the circuit rather than the actual transistor-by-transistor operation of the circuit. Simulation results using macromodels can be quite accurate. Generally, the more complex the macromodel, the more accurate the results.

Be sure you know the limitations of the macromodels you use in your simulations. Do not take them at face value. They may not model some circuit functions that are critical to your particular application. Characteristics such as common-mode response, output swing limitations, and performance under certain fault conditions may not be accurately modeled.

Comparators

Comparators are very similar to op amps except they have no compensation capacitor and a specialized output stage. Comparators are designed to provide a digital output that indicates the relative amplitudes of its inputs. One of a comparator's inputs is generally used as a reference and the other used as a sense input. The comparator is basically a high-gain, open-loop amplifier that can amplify a very small voltage difference between its inputs enough to drive its output to its high or low limit.

Figure 5-65 illustrates a simple but highly useful comparator. The open collector output is very versatile and allows an easy interface with a variety of other circuits. Common-mode range is important in comparators and must include the range of voltages to be compared.

Comparators can have significant problems with signals that are very close together. Noise on the signals or within the comparator's input stage may cause the comparator to oscillate between states under these conditions. This situation usually occurs when the signal and reference are very nearly equal or when the signal is changing slowly through the transition region. The latter case occurs when the comparator is sensing a slowly charging capacitor.

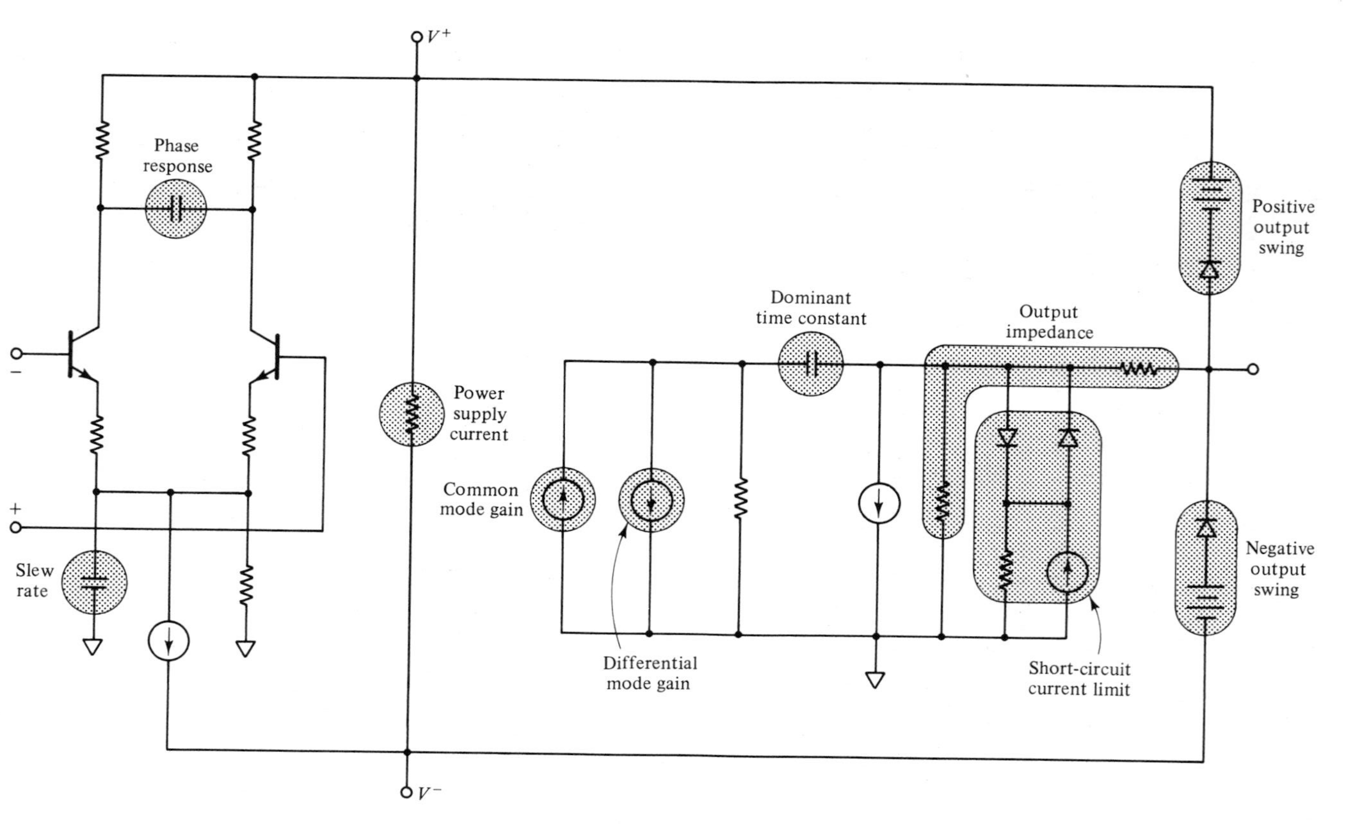

Figure 5-64 Op amp macromodel.

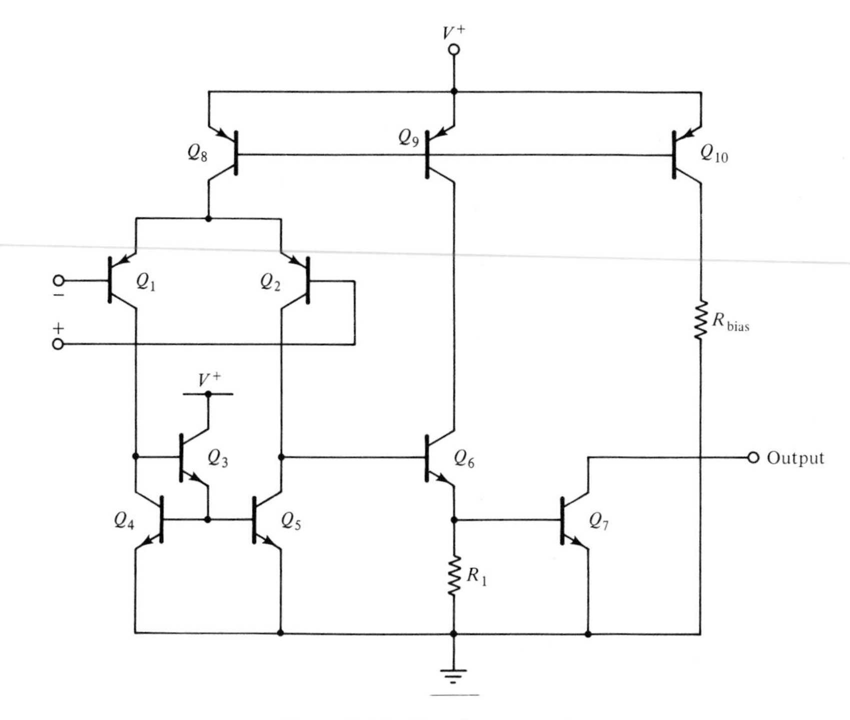

Figure 5-65 Simple comparator.

One method to address this problem is the use of hysteresis. Hysteresis constitutes positive feedback that reinforces the comparator's decision to switch. Figure 5-66 illustrates the use of hysteresis in an inverting comparator.[18]

Figure 5-67(*a*) shows the equivalent circuit with the comparator output high (output transistor off). In this case we are assuming that $R_1 \ll R_L$ and $R_1 \ll R_2$. The trip threshold V_T is found as

$$V_T = \frac{(V^+ - V_{ref})\,R_S}{R_2 + R_S} + V_{ref} = \frac{V^+R_S + V_{ref}R_2}{R_2 + R_S}$$

Figure 5-67(*b*) shows the equivalent circuit when the comparator output is low (output transistor saturated). The trip threshold V_T is found as

$$V_T = \frac{V_{ref}R_2}{R_S + R_2}$$

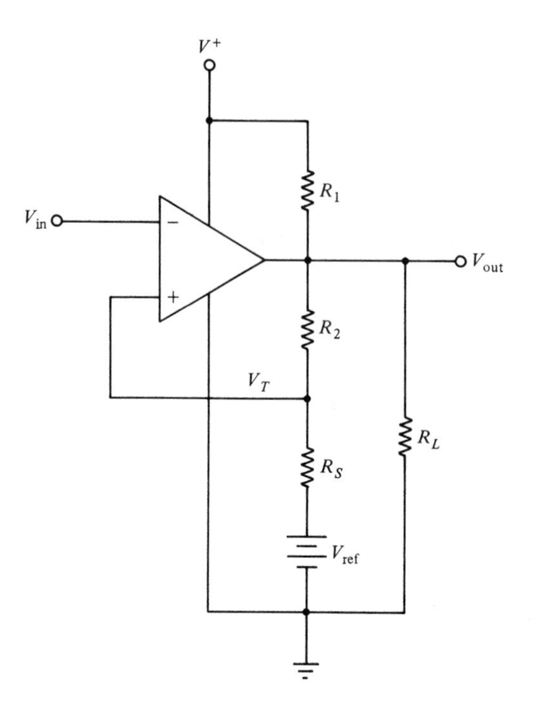

Figure 5-66 An inverting comparator with hysteresis.

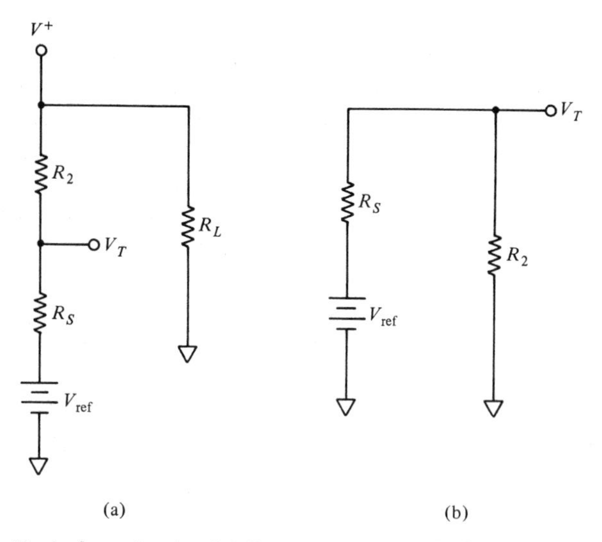

Figure 5-67 Equivalent circuits. (*a*) Comparator output high (assumes $R_1 << R_L$ and $R_1 << R_2$) and (*b*) comparator output low.

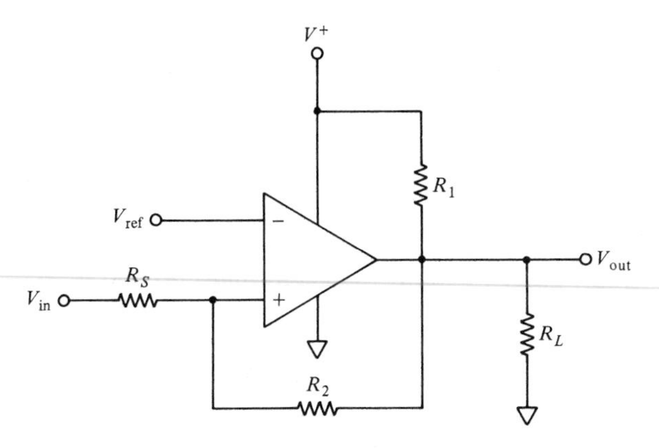

Figure 5-68 A noninverting comparator with hysteresis.

Figure 5-68 illustrates the use of hysteresis in a noninverting comparator. Figure 5-69(*a*) shows the equivalent circuit with the comparator output high (output transistor off). In this case, as before, we are assuming $R_1 << R_L$ and $R_1 << R_2$. The trip threshold V_T is found as

$$V_T = \frac{V^+ R_S + V_{in} R_2}{R_2 + R_S}$$

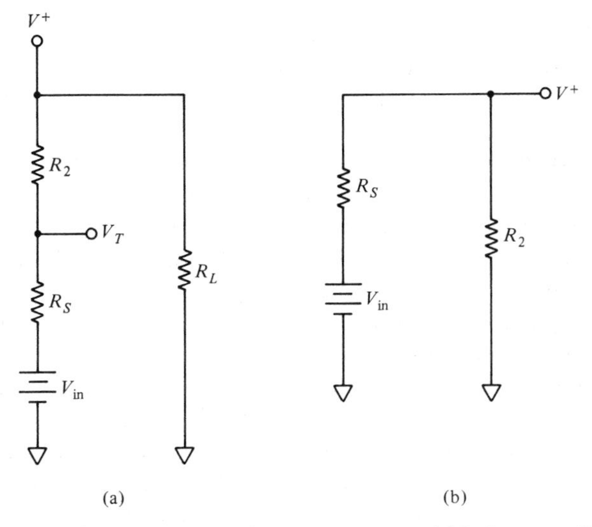

Figure 5-69 Equivalent circuits. (*a*) Comparator output high (assumes $R_1 << R_L$ and $R_1 << R_2$) and (*b*) comparator output low.

Figure 5-69(*b*) shows the equivalent circuit with the comparator output low (output transistor saturated). The trip threshold is found as

$$V_T = \frac{V_{in} R_2}{R_S + R_2}$$

In both the inverting and noninverting cases the hysteresis is the difference in the trip threshold between the output high case and output low case. The amount of hysteresis can be small or large, depending on the application. The smallest value of hysteresis that is meaningful is a value larger than the worst-case input offset that can be expected from the comparator. Typical rules of thumb for the amount of hysteresis necessary to suppress oscillations are

$$\Delta V_T = 10 \times V_{os\max}$$
$$\Delta V_T = 1\% \text{ of output swing}$$

Since the trip points can be set at virtually any value, the comparator can be used as a Schmitt trigger.

5.4 Oscillators

An oscillator is a circuit that provides an output signal with no input other than power-supply voltage. Oscillators depend on gain, gain control, positive feedback, and, to some extent, noise. It is not difficult to make a circuit oscillate if there is high enough gain. Many engineers have had major difficulties designing high-gain, wide-band amplifiers that do not oscillate. The challenge in designing an oscillator is producing an output with a known and controlled frequency of oscillation, amplitude, and waveform (spectral content) that is invariant with temperature and supply voltage changes.

The amount of precision required from the oscillator will dictate the level of circuit complexity. Oscillators with relatively modest levels of performance can be easily constructed with a single transistor and a few passive components. A high-precision oscillator may require a voltage regulator, a precision high-Q bandpass network, an AGC circuit, a buffer amplifier, and a temperature-controlled environment.

This section will review some basic oscillator circuits. These can be combined with additional circuitry, such as voltage regulators or buffer amplifiers, to obtain the desired level of performance. The references listed at the end of the chapter will provide greater detail than will be presented here.

There are many different oscillator circuits. All of them can be broken down into two main types: bandpass oscillators and time-delay oscillators. Bandpass oscillators typically output sine waves. The sine waves can be subsequently processed to yield other waveforms such as square waves, triangle waves, or sawtooth waves. The frequency-determining network for these oscillators is a frequency-selective band-pass (or band-reject) network. These are typically made of resistors and capacitors, capacitors and inductors, ceramic resonators, or quartz crystals. Some of these oscillators are tunable over wide frequency ranges while others have just enough tuning range to compensate for component variations.

Time-delay oscillators typically output square waves, exponential rising and falling waveforms, triangle waves, and sawtooth waves. The generation of sine waves by this type of oscillator usually requires waveform processing and filtering. The frequency-determining network for these oscillators is a time-delay circuit. This is frequently the charge and discharge af a capacitor or some sort of counting circuit. These oscillators can be tunable over a wide range by varying the time delay, be adjustable over a relatively narrow range, or be synchronized to some other signal.

Bandpass Oscillators

Figure 5-70 illustrates the basic circuit for a bandpass oscillator.[19] The voltage gain of the circuit is given by the following equation:

$$A_v = \frac{A_o}{1 - A_o F}$$

where

A_v = closed-loop gain
A_o = open-loop amplifier gain
F = feedback factor

Theoretically, oscillations will occur when $A_oF = 1$. In practice, A_oF must be at a level larger than unity to maintain predictable oscillations. The optimum amount of positive feedback varies with circuit implementation. Any amount of positive feedback over this optimum

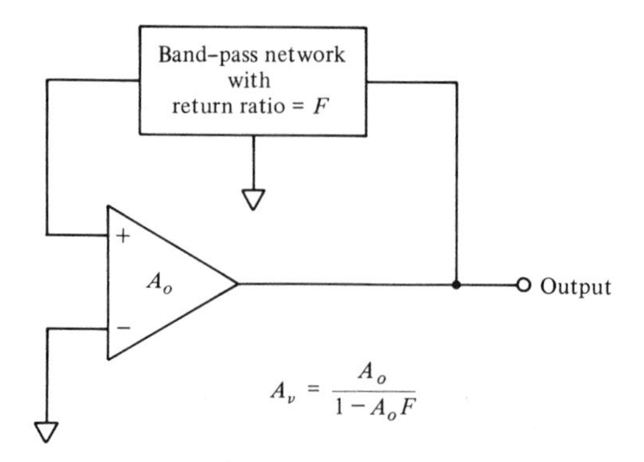

Figure 5-70 Oscillator block diagram.

amount must be regulated by a gain-controlling nonlinearity within the oscillator circuit or unpredictable performance can result. In single-transistor oscillators, this nonlinearity is the base-emitter voltage.

See Reference 11 for more detailed information on the material in this section.

LRC Oscillators

LRC resonant circuits, sometimes called tanks or resonant tank circuits, can be configured in either series or parallel, as shown in Figure 5-71. When the circuit is connected in series, impedance is minimum and current flow through the circuit is maximum at resonance. When the circuit is connected in parallel, impedance is maximum and

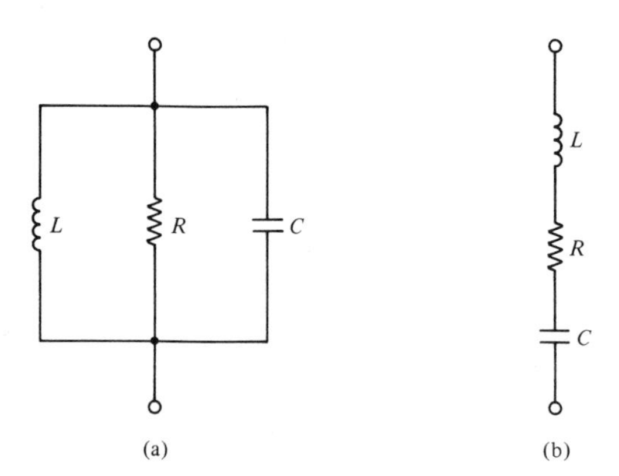

Figure 5-71 LRC resonant circuits. (*a*) Parallel and (*b*) series resonant circuits.

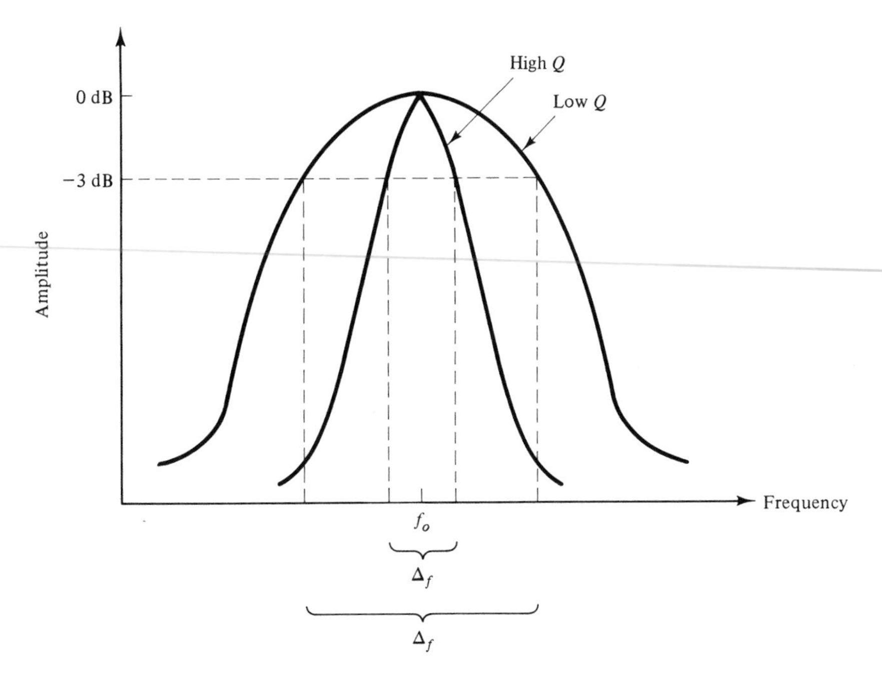

Figure 5-72 Resonant circuit response.

current flow through the circuit is minimum at resonance. Figure 5-72 illustrates the response of a low-Q and a high-Q resonant circuit. The quality factor Q of the circuit is the ratio of the energy stored in the circuit to the energy lost or dissipated. High-Q circuits have a sharp peak and a narrow bandwidth. Low-Q circuits have a more gradual peak and a wider bandwidth. Q is more easily measured by the ratio of the resonant frequency f_o to the 3-dB bandwidth.

$$Q = \frac{f_o}{\Delta f}$$

Obviously, the Q of the resonant circuit will have an effect on the return ratio versus frequency and therefore the stability of the oscillator. The dissipating element that affects the Q is the resistance. The resistance is included two ways. A resistor can be purposely added to lower the Q. Most often, the effective resistance is due either to the parasitic resistance in the wire used to wind the inductor or the effects of core loss. Remember, the resistance is at the frequency of interest. Using wire

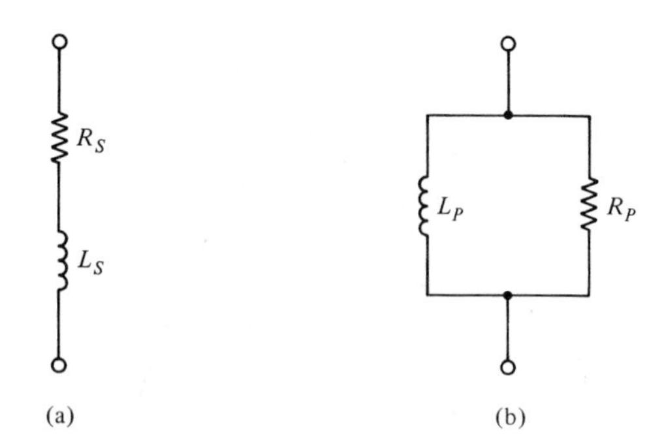

Figure 5-73 Series-to-parallel conversion. (*a*) Equivalent series LR and (*b*) equivalent parallel LR.

with a larger effective cross section will reduce the parasitic resistance. This means large wire at low frequencies or litz wire at higher frequencies. Core material choice will also impact achievable circuit *Q*.

The ohmic resistance of an inductor is easy to measure. This resistance appears in series with the inductor. The series inductance L_s and the series resistance R_s can be converted into an equivalent parallel inductance L_p and an equivalent parallel resistance R_p as shown in Figure 5-73 by the following equations:[20]

$$Q = \frac{2\pi f L_s}{R_s}$$

$$R_P = R_S(1 + Q_2)$$

$$L_P = L_S\frac{(1 + Q_2)}{Q_2}$$

If L_p and R_p are known, the above equations can be solved for L_s and R_s. In either the series or parallel resonant mode, the resonant frequency is

$$f_o = \frac{1}{2\pi\sqrt{LC}}$$

Figures 5-74 through Figure 5-77 illustrate typical oscillator circuits using LRC resonant circuits.[21] The Colpitts, Hartley, and tuned collector (also called tickler) oscillator circuits are classic configurations originally developed for vacuum tubes. The tuned collector oscillator in Figure 5-77 is well suited for ASIC design since only one pin is required

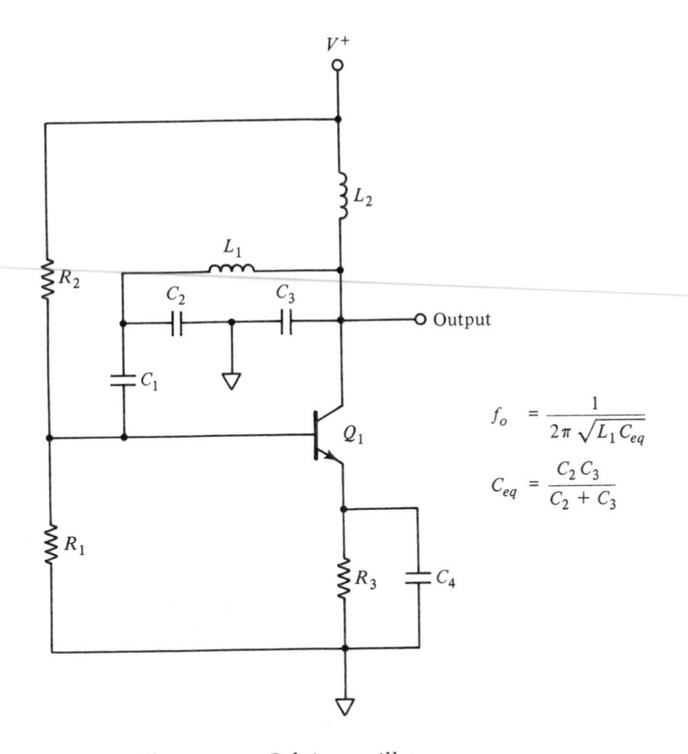

Figure 5-74 Colpitts oscillator.

for the resonant circuit and output. An emitter-follower buffer can be connected to the output to isolate a low impedance load.

Crystal Oscillators

A crystal can be viewed as a very high-Q LRC circuit. The equivalent circuit of a crystal is shown in Figure 5-78.[22,23] Crystals are usually operated in the series resonant mode:

$$f_s = \frac{1}{2\pi\sqrt{LC}}$$

where

f_s = series resonant frequency
L = motional arm inductance
C = motional arm capacitance

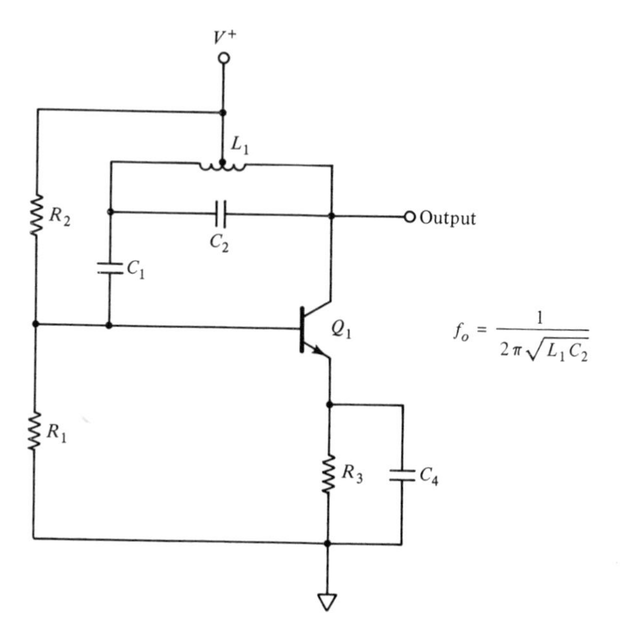

Figure 5-75 Hartley oscillator.

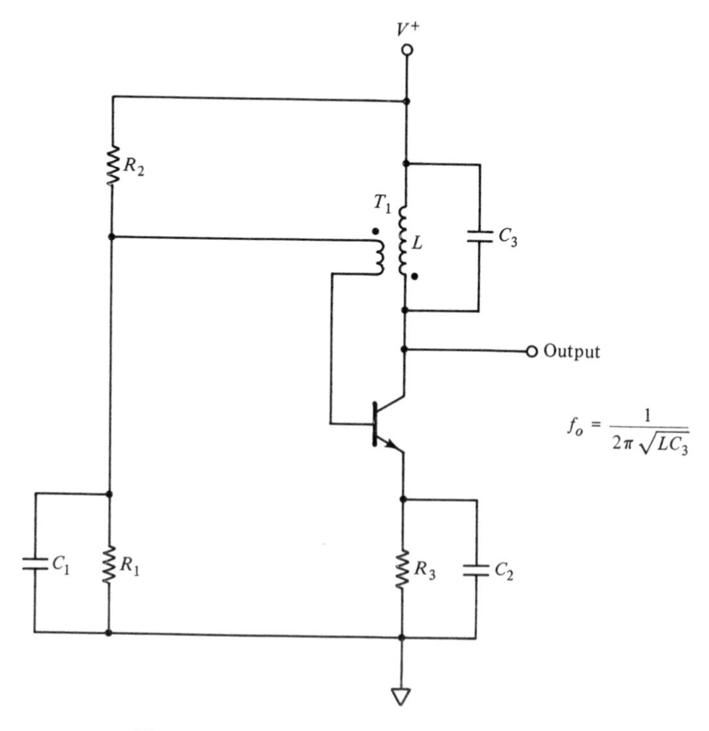

Figure 5-76 Tuned collector/tickler oscillator.

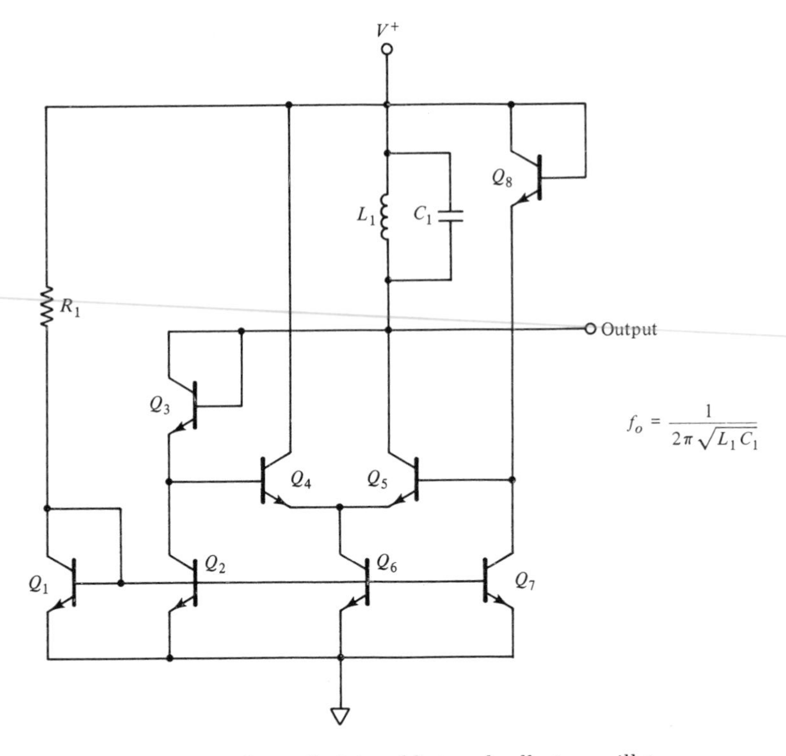

Figure 5-77 An easily integrable-tuned collector oscillator.

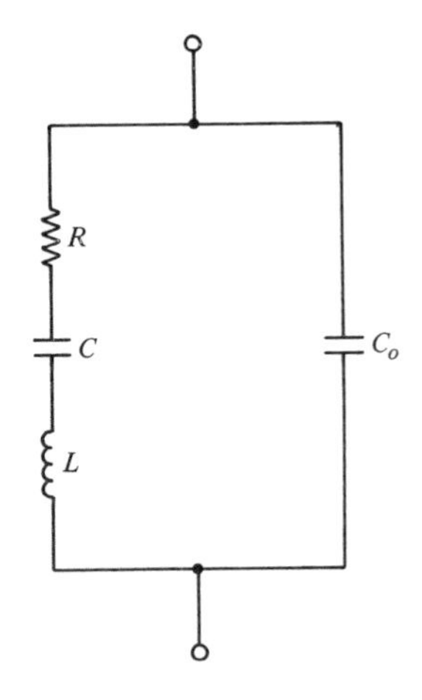

Figure 5-78 Crystal equivalent circuit.

The crystal can also resonate in the parallel mode with its holder capacitance C_o at the frequency

$$f = f_s\left(1 + \frac{C}{2C_o}\right)$$

or with an external parallel capacitor C_p at the frequency

$$f \doteq f_s\left(1 + \frac{C}{2C_o + C_p}\right)$$

Crystals with an AT cut can be operated on their third or fifth mechanical overtone. This technique is generally used above 20 MHz. Oscillators using crystals in this mode must have additional filtering to ensure that the crystal is operating on the correct overtone.

Figures 5-79 and 5-80 illustrate typical crystal oscillator circuits. In each case, L_2 is tuned to resonate with C_o to ensure operation at the proper crystal overtone.[24]

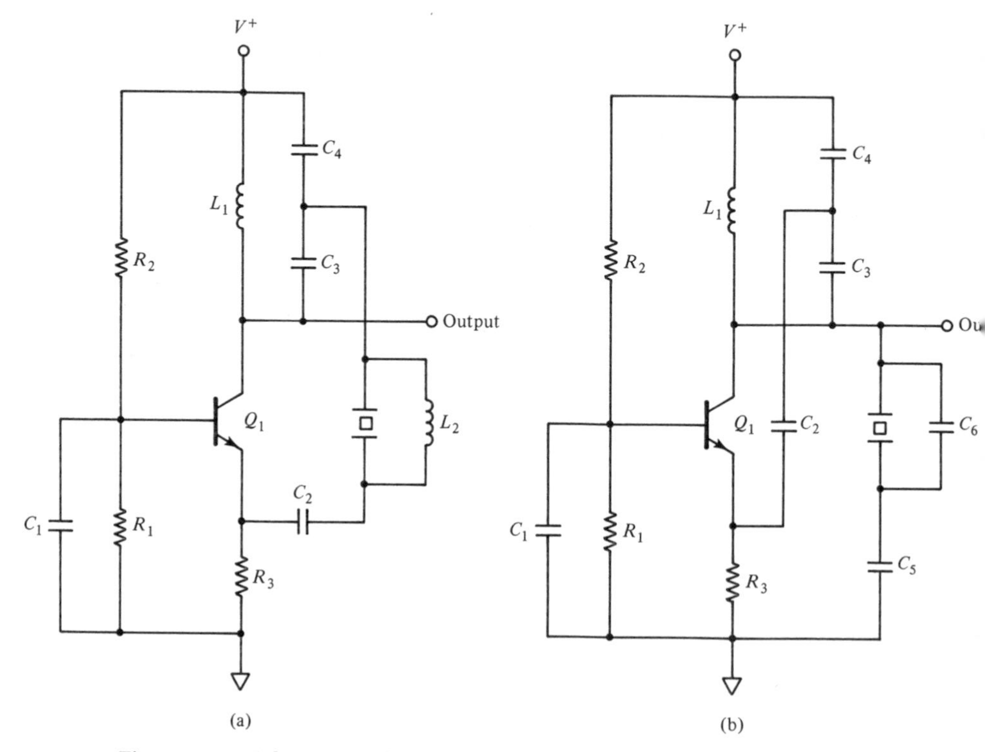

Figure 5-79 Colpitts crystal oscillator. (a) Series resonant and (b) parallel resonant.

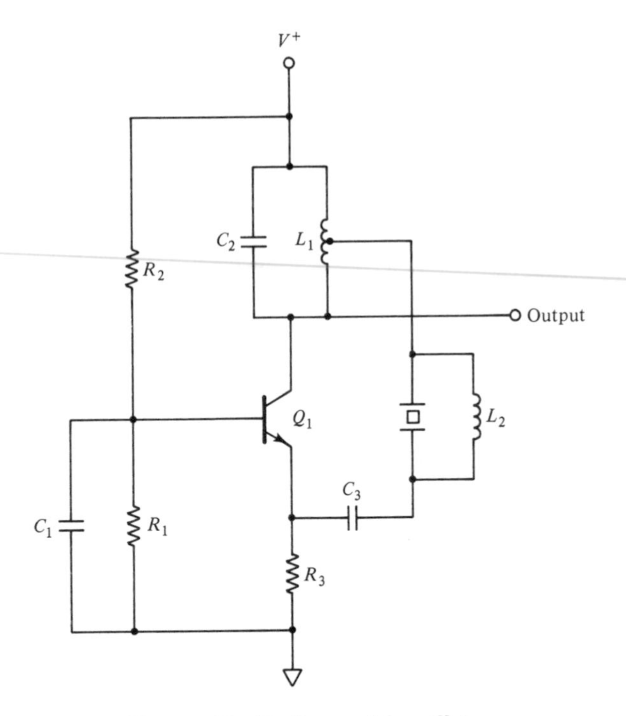

Figure 5-80 Hartley crystal oscillator.

RC *Oscillators*

RC oscillators operate at low fixed frequencies. They require multiple element values to be changed to change frequency. This makes it impractical to operate these oscillators over a widely adjustable range of frequencies. Typically, the upper frequency limit is about 1 MHz.

Phase-Shift Oscillator

A classical phase-shift oscillator is illustrated in Figure 5-81.[25,26] Each section of the cascaded *RC* network provides 60° of phase shift to the signal being fed back. The 180° of phase shift through the *RC* network plus the 180° phase shift through the amplifier yield the necessary 360° of total phase shift necessary for positive feedback. For simplicity, all of the capacitors are chosen to be equal and $R = R_1 // R_2$. The gain is set by R_f. The frequency of oscillation is given by

$$f_o = \frac{1}{\sqrt{6}RC}$$

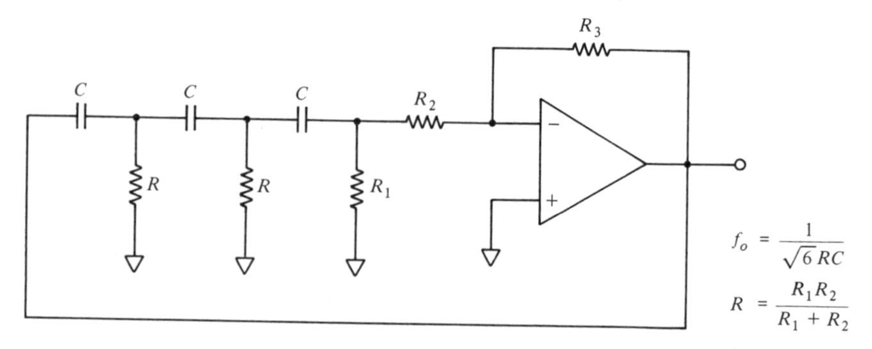

Figure 5-81 Phase-shift oscillator.

Wien Bridge Oscillator

A Wien bridge oscillator configuration is shown in Figure 5-82.[27] The frequency of oscillation is

$$f_o = \frac{1}{2\pi RC}$$

The minimum gain required to sustain oscillations is 3. Slightly more gain will be necessary to ensure that the oscillation is self-starting. A gain control network that modifies the effective value of R_1 or R_2 to

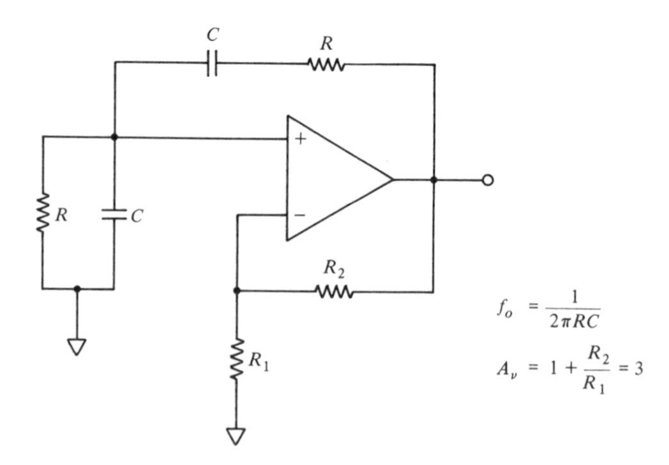

Figure 5-82 Wien bridge oscillator.

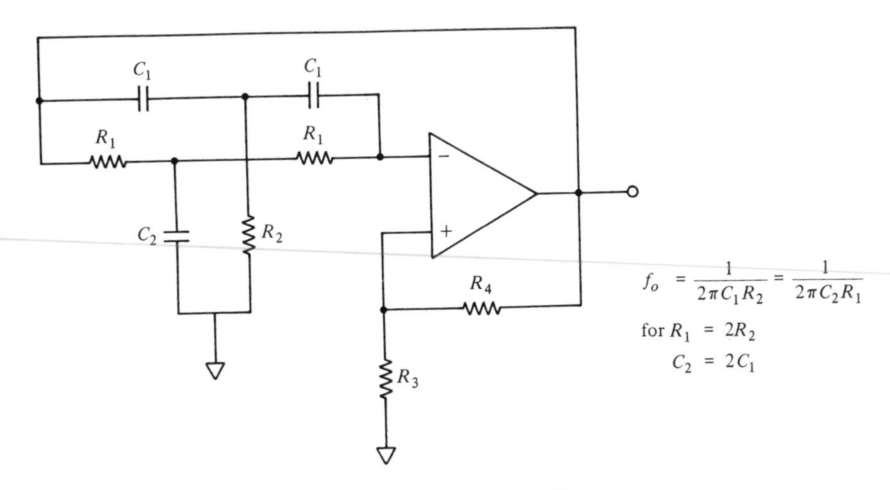

Figure 5-83 Twin-tee oscillator.

maintain the output amplitude at a predetermined value will improve waveform distortion and output amplitude stability.

Twin-Tee Oscillator

A twin-tee oscillator configuration is illustrated in Figure 5-83.[28,29] The twin-tee provides a null (notch) at the frequency

$$f_o = \frac{1}{2\pi C_1 R_2} = \frac{1}{2\pi C_2 R_1}$$

At the null frequency, negative feedback ceases and the circuit oscillates. The positive feedback is determined by R_3 and R_4.

Time-Delay Oscillators

Time-delay oscillators rely on the charging and discharging of a capacitor and a comparator to sense the capacitor voltage and control the charge and discharge of the capacitor.[30] The trip points for the comparator are set with some type of hysteresis, which defines the charge and discharge range of the capacitor. These circuits can be designed to oscillate at frequencies that are a fraction of a cycle per second, to frequencies over 50 MHz. Relatively low frequencies can be generated with relatively small-value capacitors by using one of these oscillators

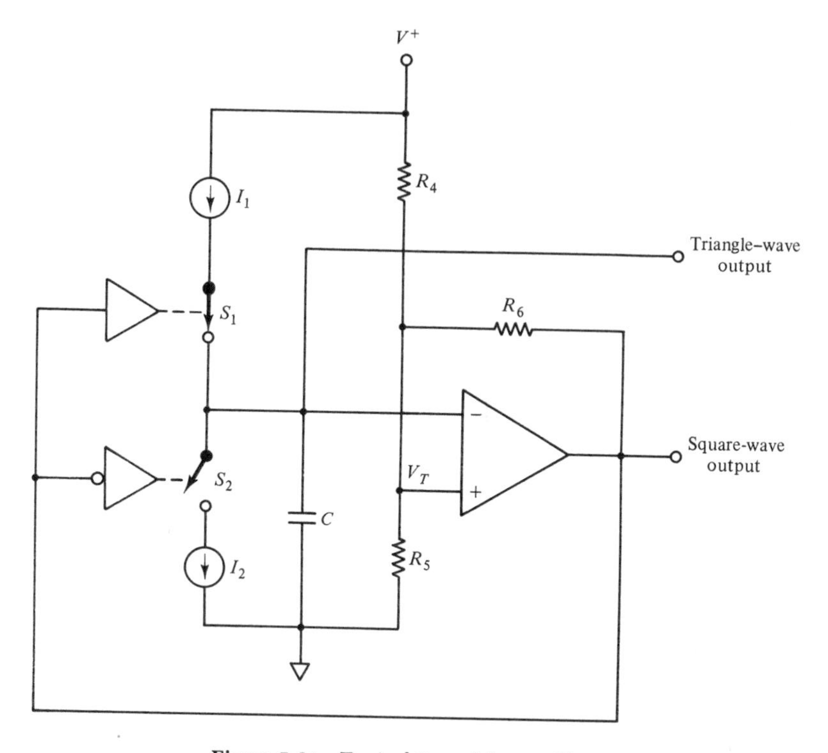

Figure 5-84 Typical time-delay oscillator.

in conjunction with a counter. The outputs are normally square waves or triangle waves.

Figure 5-84 illustrates the basic operation of these circuits. Upon start-up, the capacitor is discharged, the comparator output is high (at V^+), S_1 is closed, and the comparator threshold is

$$V_T^+ = \frac{V^+ R_5}{(R_4 // R_6) + R_5}$$

The capacitor charges linearly at the rate

$$\frac{dV}{dt} = \frac{I_1}{C}$$

until V_T is reached. At this point, the comparator output switches low, S_1 opens, S_2 closes and the new V_T is

$$V_T = \frac{V^+ (R_5 // R_6)}{(R_5 // R_6) + R_4}$$

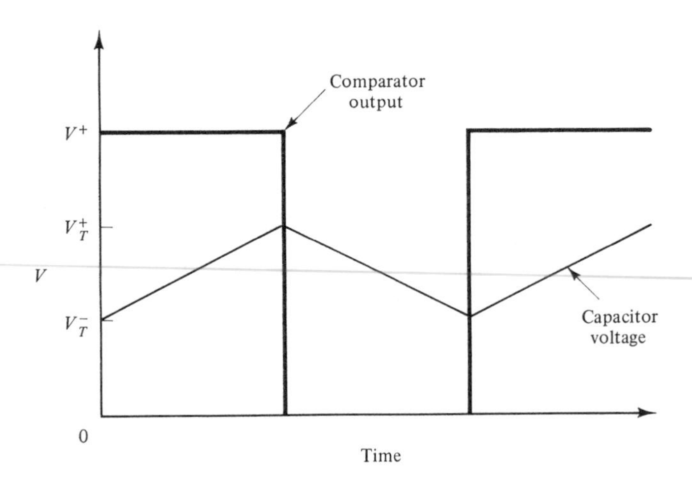

Figure 5-85 Oscillator output waveforms.

The capacitor now discharges linearly at the rate

$$\frac{dV}{dt} = \frac{I_2}{C}$$

until V_T is reached, at which point the comparator output switches high and the process is repeated.

The voltage on the capacitor is a triangle wave due to the linear charge and discharge of the capacitor. The comparator output is a square wave that switches low at the triangle wave's highest point and switches high at the triangle wave's lowest point, as illustrated in Figure 5-85. The duty cycle of the oscillation is 50% if $I_1 = I_2$. The duty cycle can be easily modified by changing the relative values of I_1 and I_2. A sawtooth waveform could be generated if S_2 were connected to ground instead of I_2.

A practical integrated implementation of this circuit is shown in Figure 5-86. Q_1 and R_1 determine the currents for the circuit. By decreasing R_1 and increasing the charge and discharge currents, the frequency of oscillation can be adjusted. Q_5 provides a charge current for the capacitor equal to I_1. Q_2 and Q_3 provide a current equal to twice I_1, which is mirrored by Q_4 and Q_7, to the emitters of the differential pair Q_6 and Q_8. R_2 and R_3 set the base voltage of Q_6 at a point midway in the charge and discharge range of the capacitor. Q_6 and Q_8 form a current switch to charge and discharge the capacitor C. When the base of Q_8 is at the top of the capacitor's charge range, Q_6 is turned off, I_1 from Q_5

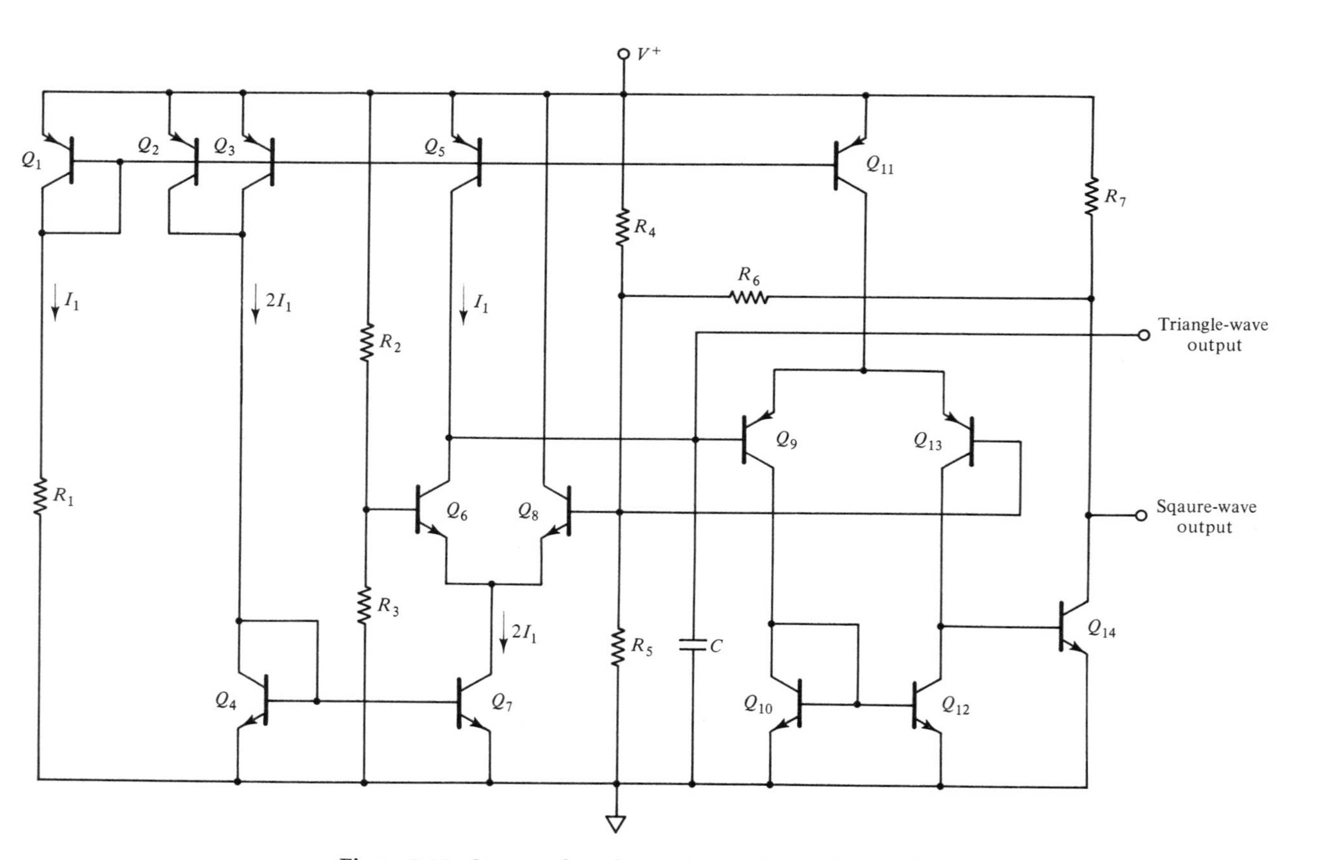

Figure 5-86 Integrated implementation of time-delay oscillator.

charges the capacitor, and Q_8 supplies $2I_1$ to the collector of Q_7. Q_9, Q_{10}, Q_{11}, Q_{12}, Q_{13}, and Q_{14} form a comparator. The base of Q_9 senses the capacitor voltage and the base of Q_{13} senses V_T. While the capacitor is charging, the base of Q_{13} is at a higher potential than the base of Q_9. Therefore, Q_{13} is off. The collector current of Q_{11} is conducted through Q_9 to Q_{10} and mirrored to the collector of Q_{12}. Since Q_{13} is off, Q_{12} saturates and pulls the base of Q_{14} low, turning off Q_{14} and placing R_7 and R_6 in parallel with R_4 and the voltage on the base of Q_{13} at the highest potential of the capacitor charge cycle. When the capacitor charges to this potential, Q_9, Q_{10}, and Q_{12} are turned off, Q_{13} turns on, supplying base current to Q_{14}, which then saturates, placing V_t^- on the base of Q_{13}, turning off Q_8 and turning on Q_6. Q_6 provides $2I_1$ to the collector of Q_7 by conducting I_1 from the capacitor and I_1 from the collector of Q_5. Thus, the charge and discharge current to the capacitor is equal to I_1.

5.5 Nonlinear Circuits

The inherent nonlinearity of *pn* junctions gives rise to a rich variety of circuit functions and applications. These include

- Multiplication
- Modulation
- Demodulation
- Phase discrimination
- Square root
- Vector magnitude
- Log amplifiers
- Antilog amplifiers

To illustrate some basic principles, we will analyze the dc transfer characteristics of an emitter-coupled differential pair. To simplify the analysis and avoid becoming lost in unnecessary detail, we will assume a very high beta and neglect other parasitic effects such as transistor output impedance.

Figure 5-87(*a*) illustrates a basic current source-biased emitter-coupled pair. Figure 5-87(*b*) illustrates the response of this circuit. From Figure 5-87(*a*) it can be seen that the following relationships hold:

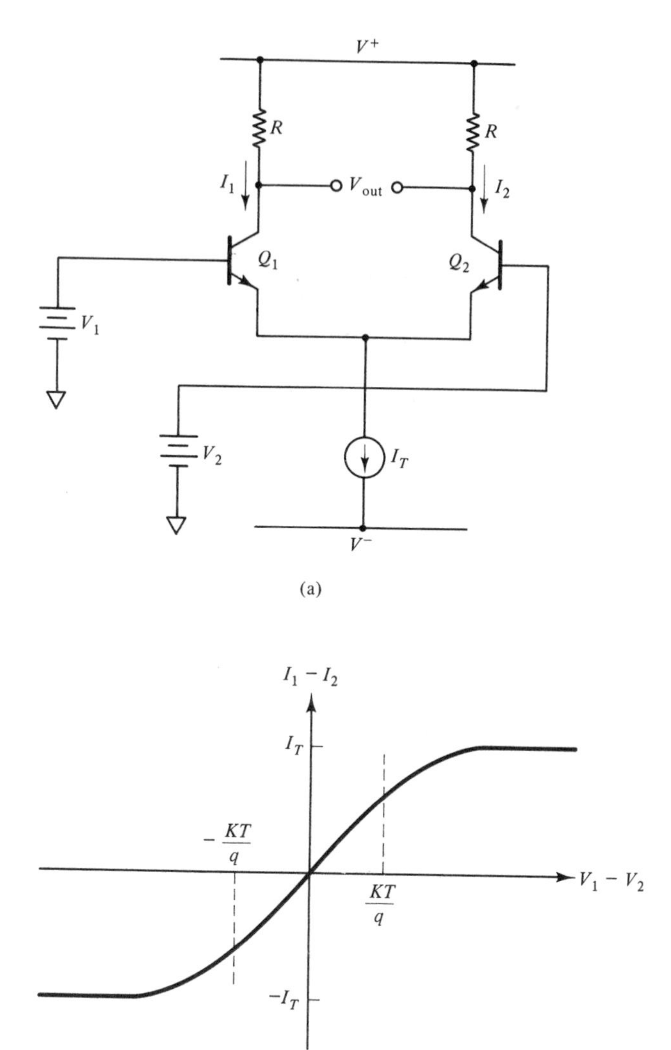

Figure 5-87 Emitter-coupled differential pair. (*a*) Circuit and (*b*) response.

$$0 = V_1 - V_{be1} + V_{be2} - V_2$$

$$V_{be1} = \frac{kT}{q}\ln\left(\frac{I_1}{I_s}\right)$$

$$V_{be2} = \frac{kT}{q}\ln\left(\frac{I_2}{I_s}\right)$$

$$V_{be1} - V_{be2} = \frac{kT}{q}\ln\left(\frac{I_1}{I_2}\right) = \Delta V_{be} = V_{in}$$

$$\frac{I_1}{I_2} = \exp\left(\frac{q\Delta V_{be}}{kT}\right)$$

$$I_1 = I_2\exp\left(\frac{q\Delta V_{be}}{kT}\right)$$

Also,

$$I_T = I_1 + I_2$$

$$I_2 = I_T - I_1$$

$$I_2 = I_T - I_2\exp\left(\frac{q\Delta V_{be}}{kT}\right)$$

$$I_T = I_2\left(1 + \exp\left(\frac{q\Delta V_{be}}{kT}\right)\right)$$

$$I_2 = \frac{I_T}{1 + \exp\left(\frac{q\Delta V_{be}}{kT}\right)}$$

$$I_1 = I_T - I_2$$

$$I_1 = I_T - \frac{I_T}{1 + \exp\left(\frac{q\Delta V_{be}}{kT}\right)}$$

$$I_1 = I_T\left(\frac{1 + \exp\left(\frac{q\Delta V_{be}}{kT}\right) - 1}{1 + \exp\left(\frac{q\Delta V_{be}}{kT}\right)}\right)$$

$$I_1 = \frac{I_T}{1 + \exp\left(\frac{-q\Delta V_{be}}{kT}\right)}$$

The difference in collector currents can be found as

$$\Delta I = I_1 - I_2$$

$$\Delta I = \frac{I_T}{1 + \exp\left(\frac{-q\Delta V_{be}}{kT}\right)} - \frac{I_T}{1 + \exp\left(\frac{q\Delta V_{be}}{kT}\right)}$$

For simplicity, let

$$e^x = \exp\left(\frac{q\Delta V_{be}}{kT}\right)$$

Substituting, this yields

$$\Delta I = I_T\left(\frac{e^x - e^x}{2 + e^x + e^x}\right) = I_T\left(\frac{\frac{1}{2}(e^x - e^x)}{1 + \frac{1}{2}(e^x + e^x)}\right)$$

$$\Delta I = I_T\left(\frac{\sinh x}{1 + \cosh x}\right) = I_T \tanh \frac{x}{2}$$

$$\Delta I = I_T \tanh \frac{q\Delta V_{be}}{2kT}$$

The output voltage is

$$V_{out} = \Delta IR = RI_T \tanh \frac{q\Delta V_{be}}{2kT}$$

For

$$\Delta V_{be} \ll \frac{2kT}{q}$$

$$\tanh \frac{q\Delta V_{be}}{2kT} \approx \frac{q\Delta V_{be}}{2kT}$$

$$V_{out} = RI_T \frac{q\Delta V_{be}}{2kT}$$

Now, if we make I_T proportional to a voltage, as shown in Figure 5-88, we can create a two-quadrant multiplier. For this circuit we will assume

$$R_b \gg \frac{kT}{qI_T}$$

$$V_{in2} \gg V_{be3}$$

$$I_T \approx \frac{V_{in2}}{R_b}$$

$$V_{in1} < \frac{2kT}{q}$$

$$V_{out} = RI_T \frac{q\Delta V_{be}}{2kT}$$

$$V_{out} = \frac{R_C V_{in2}}{R_b} \frac{qV_{in1}}{kT} = \left(\frac{R_C q}{R_b kT}\right) V_{in1} V_{in2}$$

The useful input voltage range of this circuit can be extended by predistorting V_{in1} with an inverse hyperbolic tangent function to cancel the hyperbolic tangent response of the differential pair. This can be accomplished as illustrated in Figure 5-89.

The value for I_b is the quiescent bias current through the diode-connected transistors Q_1 and Q_2. The signal current is $g_m V_{in}$, which is added to or subtracted from the bias current. The response of the circuit is calculated as follows:

$$V_{out} = V_{be1} - V_{be2}$$

$$V_{out} = \frac{kT}{q} \ln \frac{I_1}{I_2} = \frac{kT}{q} \ln \frac{I_b + g_m V_{in}}{I_b - g_m V_{in}}$$

$$V_{out} = \frac{kT}{q} \ln \frac{1 + (g_m V_{in}/I_b)}{1 - (g_m V_{in}/I_b)}$$

Since

$$\tanh^{-1} x = \frac{1}{2} \ln \left(\frac{1 + x}{1 - x}\right)$$

$$V_{out} = \frac{2kT}{q} \tanh^{-1} \left(\frac{g_m V_{in}}{I_b}\right)$$

Where g_m is the transconductance of the voltage to current converter.

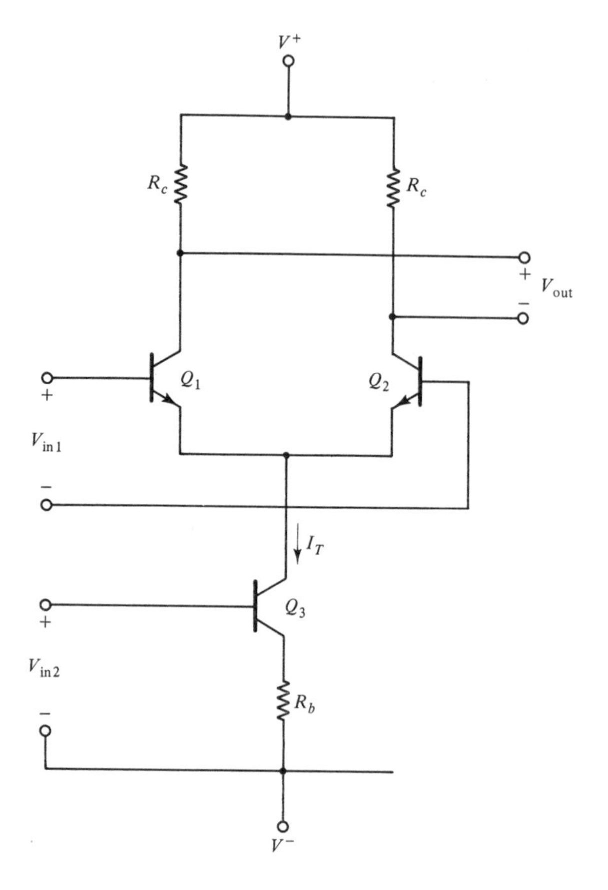

Figure 5-88 Basic two-quadrant multiplier.

The inverse hyperbolic tangent predistortion circuit can be combined with the two-quadrant multiplier in Figure 5-88 to yield the improved two-quadrant multiplier illustrated in Figure 5-90. A second two-quadrant multiplier can be added to the circuit in Figure 5-90 in a cross-coupled fashion to yield a four quadrant multiplier as illustrated in Figure 5-90. Inverse hyperbolic tangent predistortion circuits could also be added to the input ports of this circuit to improve its input dynamic range.

The four-quadrant multiplier is useful in the several different applications that follow.

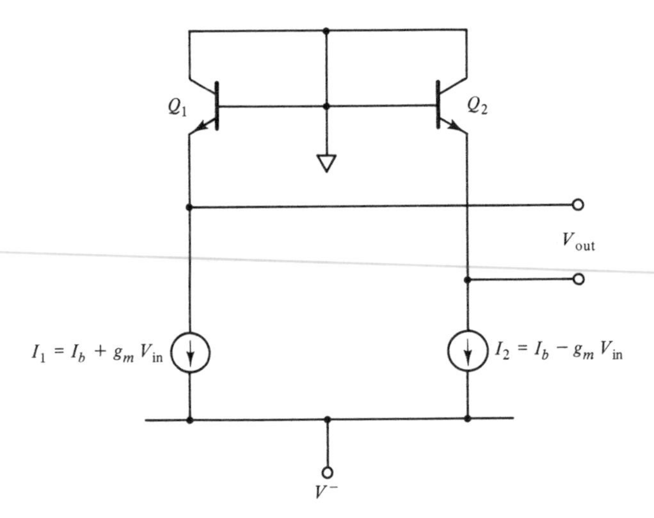

Figure 5-89 Tanh^{-1} predistortion circuit.

- Multiplier

$$V_{in1} < \frac{2kT}{q} \text{ (or } \tanh^{-1} \text{ predistortion used)}$$

$$V_{in2} < \frac{2kT}{q} \text{ (or } \tanh^{-1} \text{ predistortion used)}$$

- Modulator/Demodulator

$$V_{in2} \gg \frac{2kT}{q} \text{ (no predistortion used)}$$

$$V_{in2} < \frac{2kT}{q} \text{ (or } \tanh^{-1} \text{ predistortion used)}$$

- Phase Detector

$$V_{in1} \gg \frac{2kT}{q} \text{ (no predistortion used)}$$

$$V_{in2} \gg \frac{2kT}{q} \text{ (no predistortion used)}$$

where V_{in1} and V_{in2} are at the same frequency.

For more details on the use of multipliers see References 31 and 32.

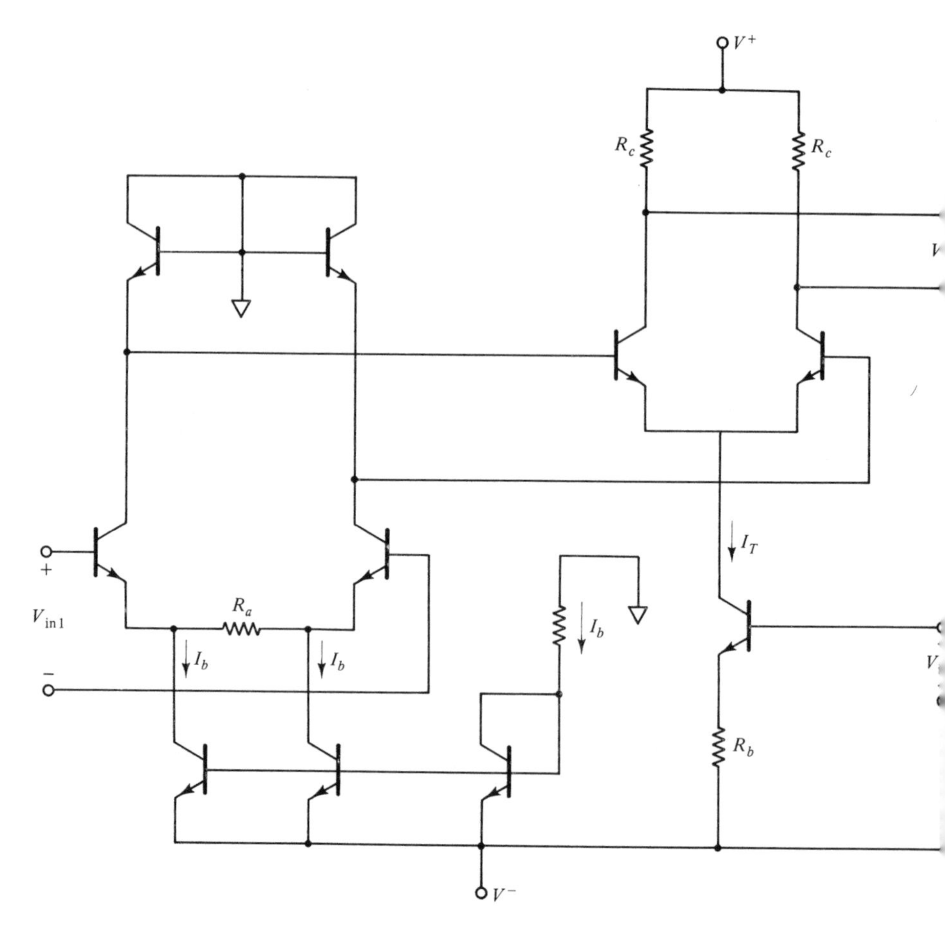

Figure 5-90 An improved two-quadrant multiplier.

Many other nonlinear functions can be performed using combinations of transistors and current sources. The circuit in Figure 5-92, for example, outputs the square root of two input currents I_1 and I_2.[33–36] Assumptions:

beta very large
matched (geometrically and thermally) transistors
parasitic effects neglected

$$V_{be2} + V_{be1} = V_{be3} + V_{be4}$$

$$\frac{kT}{q}\ln\frac{I_1}{I_S} + \frac{kT}{q}\ln\frac{I_2}{I_S} = \frac{kT}{q}\ln\frac{I_{out}}{I_S} + \frac{kT}{q}\ln\frac{I_{out}}{I_S}$$

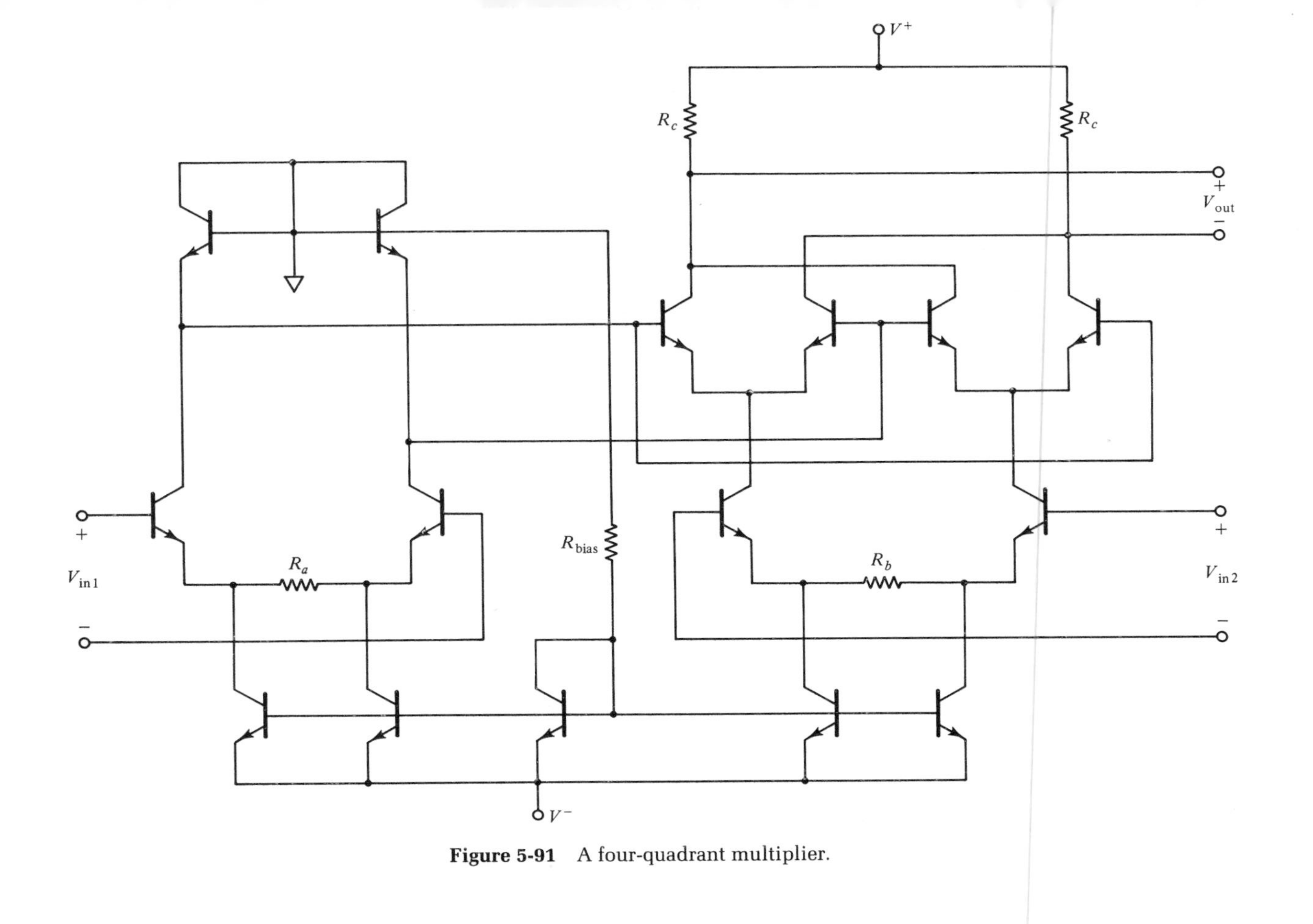

Figure 5-91 A four-quadrant multiplier.

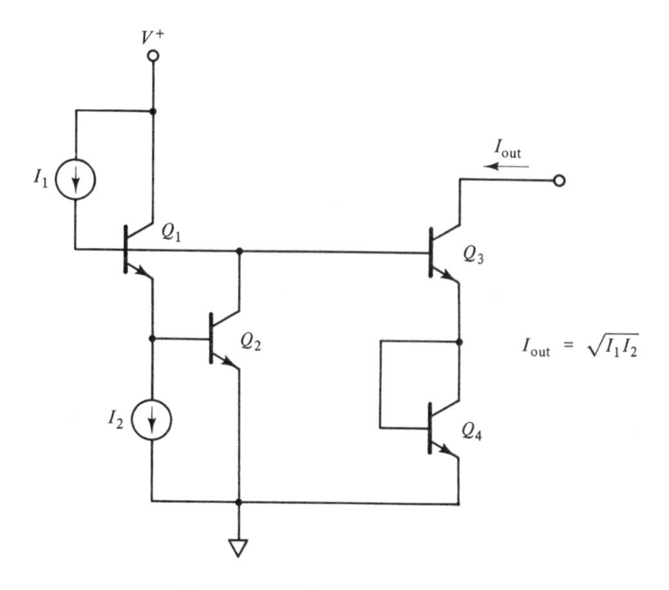

Figure 5-92 Square-root circuit.

Simplifying,

$$\ln \frac{I_1 I_2}{I_S^2} = \ln \frac{I_{out}^2}{I_S^2}$$

$$I_1 I_2 = I_{out}^2$$

$$I_{out} = \sqrt{I_1 I_2}$$

A similar circuit can be used to find the vector magnitude of the two input currents I_1 and I_2.[33–36] (Figure 5-93.)

Assumptions:

beta very large
matched (geometrically and thermally) transistors
parasitic effects neglected
an equal current split between Q_1, Q_2 and Q_3, Q_4

$$V_{be1} + V_{be3} = V_{be5} + V_{be6}$$

$$\frac{kT}{q} \ln \frac{I_1}{2I_S} + \frac{kT}{q} \ln \frac{I_1}{2I_S} = \frac{kT}{q} \ln\left(\frac{I_2}{I_S} + \frac{I_{out} - I_2}{2I_S}\right) + \frac{kT}{q} \ln \frac{I_{out} - I_2}{2I_S}$$

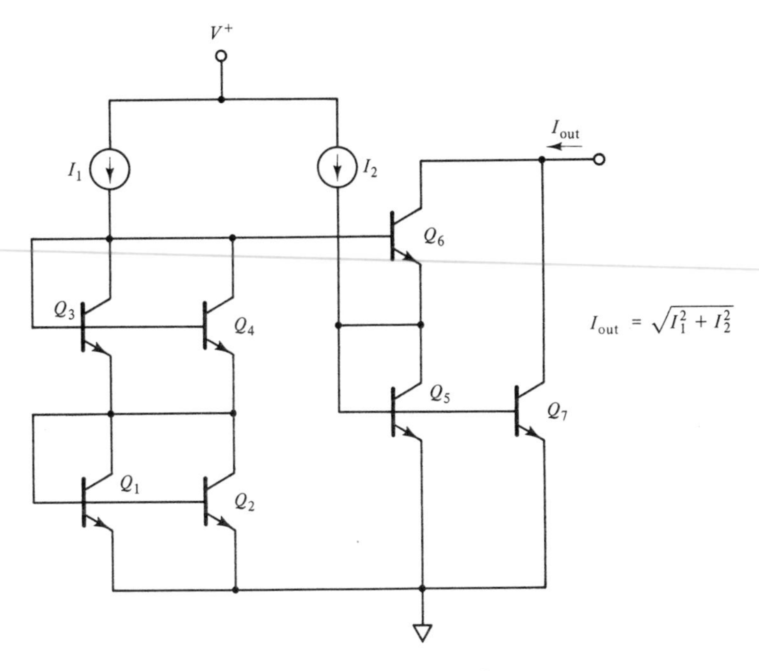

Figure 5-93 Vector magnitude circuit.

Simplifying,

$$\ln \frac{I_1^2}{4I_S^2} = \ln \frac{(I_{out} + I_2) + (I_{out} - I_2)}{4I_S^2}$$

$$I_1^2 = I_{out}^2 - I_2^2$$

$$I_{out} = \sqrt{I_1^2 + I_2^2}$$

The circuit in Figure 5-94 is a basic log amplifier.[37] The V_{be} of Q_1 is used to determine the amplifier's output voltage in response to the input current.

$$I = \frac{V_{in}}{R}$$

$$V_{out} = V_{be1} = -\frac{kT}{q} \ln \frac{V_{in}}{RI_S}$$

The circuit in Figure 5-95 has the role of the V_{be} and the resistor reversed. The input current responds exponentially to the input voltage

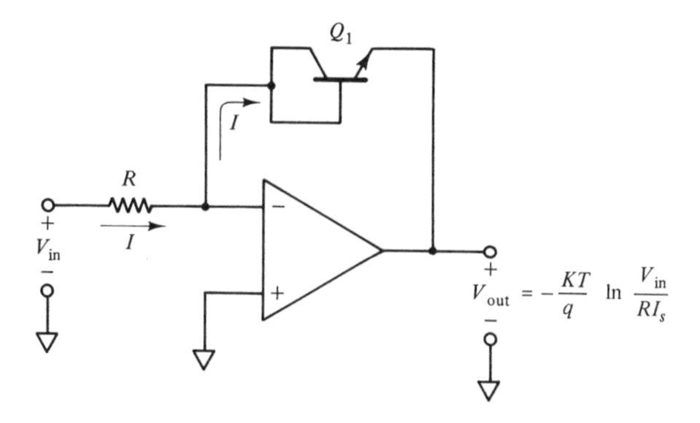

Figure 5-94 Basic log amplifier.

V_{in}. The resistor converts this exponential current into an exponential voltage.

$$I = I_S \exp\left(\frac{qV_{in}}{kT}\right)$$

$$V_{out} = -RI_S \exp\left(\frac{qV_{in}}{kT}\right)$$

The preceding examples are a few of the wide variety of nonlinear circuits possible using the inherent characteristics of *pn* junctions.

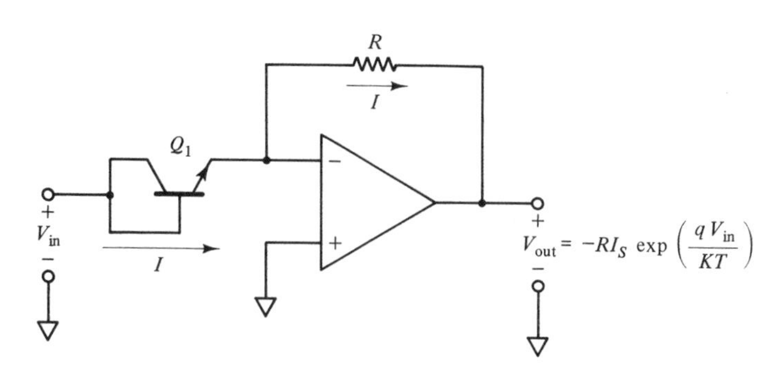

Figure 5-95 Basic antilog amplifier.

5.6 Timers

Timers are very useful in many applications. They can have short or very long delays. They can be used for timing a circuit function such as sounding an alarm for a fixed period of time when certain conditions are present. Timers can delay circuit operation for a period of time to allow the circuit to stabilize after power-up. Timer signals can modulate other signals off or on. They can provide a power-down function if a circuit has not been activated for a period of time. This last application can be very useful in battery-powered devices or as an energy saving feature.

Timers universally use the charge and discharge of a capacitor to generate the time delay.[30] A typical timer, shown in Figure 5-96, has a capacitor, a method for charging and discharging the capacitor, a reference, and a comparator. An oscillator and a counter chain can also be used to generate a long-interval timer. This implementation has the advantage of trading a large-value (large-size) capacitor for a silicon area in the form of the counter.

Since the time delay depends on charging (or discharging) a capacitor, the length of the delay depends on the size of the capacitor and/or the amount of charging current available. There are two methods for charging a capacitor: a resistor connected to a voltage or a current

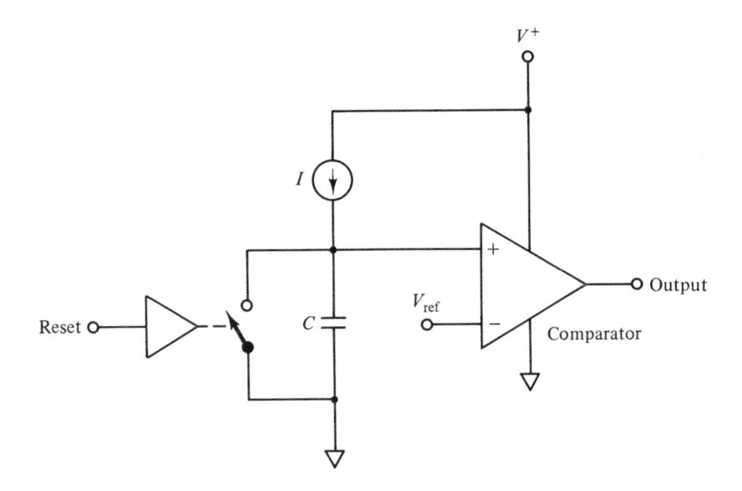

Figure 5-96 Typical timer.

source. For a given-size capacitor, the time delay is limited by the capacitor type and its inherent leakages. A large-value foil electrolytic capacitor (500–1000 μF) can have a leakage current of 400–500 nA. A solid tantalum capacitor will have far lower leakages. Polystyrene and polycarbonate capacitors will have very low leakages. Some types of capacitors have very wide tolerances. Ceramic capacitors can vary +80 to −20% from unit to unit. This variation will obviously have a similar effect on time delays. Design for an adequate amount of adjustment in the charging circuitry if a capacitor with potentially wide tolerance is chosen. If the timing interval is critical, include variations in capacitor value with temperature changes and comparator bias current in your error analysis. These considerations are important for very long time timers.

Resistive Charging Circuit

A typical RC charging circuit used in timers is shown in Figure 5-97.[30] Assuming that the capacitor voltage V_C starts at 0, the voltage on the capacitor is determined by the following equation:

$$V_C = V^+\left(1 - \exp\left(-\frac{t}{\tau}\right)\right)$$

where

$$\tau = RC$$
$$t = \text{time from } t = 0$$

The time for the capacitor to charge to a given V_C is

$$t = \tau \ln\left(\frac{V^+}{V^+ - V_C}\right)$$

Example Given a 100-kΩ resistor and a 12-V power supply, find the value of C to create a 1-sec delay for V_C to charge from 0 to 7 V.

$$t = (RC)\ln\left(\frac{V^+}{V^+ - V_C}\right) = (R_C)\ln\frac{12}{5} = 0.8755\ R_C$$

$$C = \frac{t}{0.8755\ R} = \frac{1}{(0.8755)100\ \text{k}\Omega} = 11.42\ \mu\text{F}$$

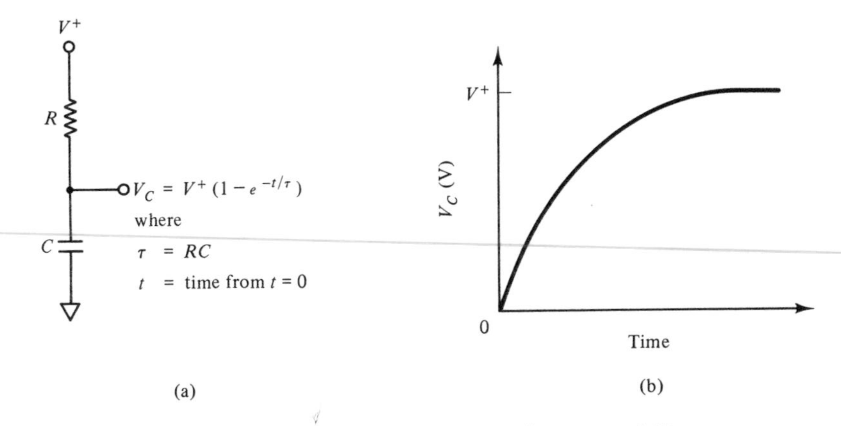

Figure 5-97 RC charging circuit. (*a*) Circuit schematic and (*b*) response.

Current-Source Charging Circuit

A current-source charging circuit is illustrated in Figure 5-98.[30] Assuming the voltage starts at 0 V, the capacitor voltage is determined by the following equation:

$$V_C = \frac{1}{C}\int_0^t i\,dt$$

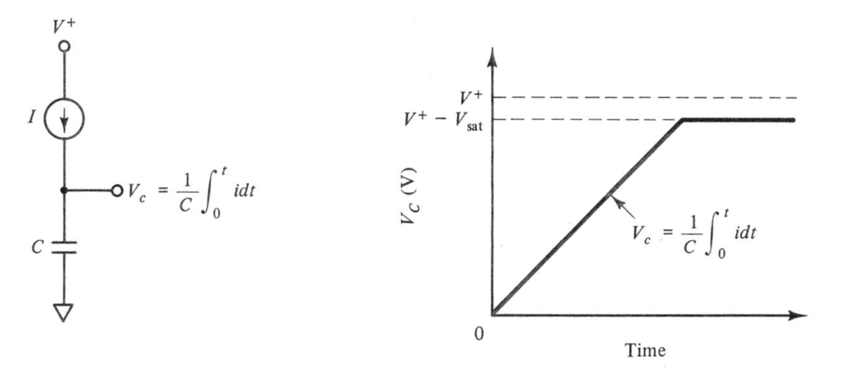

Figure 5-98 Current-source charging circuit.

If i is constant with time,

$$V_C = \frac{1}{C} it$$

where

$$t = \text{time from } t = 0$$

The time for the capacitor to charge to a given V_C is

$$t = \frac{V_C C}{i}$$

Example Given a 10-μF capacitor, what current would be necessary to charge the capacitor to 3.0 V in 1 sec?

$$t = \frac{(3.0 \text{ V})(10 \times 10^{-6})}{1} = 30\ \mu\text{A}$$

The advantage of using a current source to charge the capacitor is that V_C varies linearly with time as opposed to exponentially with an R_C arrangement.

Figure 5-99 illustrates a simple low-current timer for power-up delays. A transistor could be added to discharge the capacitor and make the timer resetable. The circuit functions as follows: When power is initially applied, the capacitor is discharged and Q_1's base is at ground potential (Q_1 is biased off). Q_2's base is at V_{ref}. Q_2 is conducting and supplies collector current to Q_3. Q_5 is cut off since no current flows through R2. Q_6 base is at ground, Q_6 is cut off, and V_{out} is pulled to V^+ through R_4. When C charges through R_1 to V_{ref}, Q_1 turns on, Q_2 turns off, and Q_5 turns on, supplying base current to Q_6, which saturates, pulling V_{out} to ground. V_{out} will remain at a logic low state until power is removed and reapplied.

Figure 5-100 is a more universal timer. It features a current-source charging circuit, a reset transistor, and an emitter follower to buffer the capacitor voltage and prevent common-mode voltage range problems for the comparator when the reset transistor is turned on.

There are a few cautions to observe when designing timer circuits. Be careful not to violate the common-mode range of the comparator.

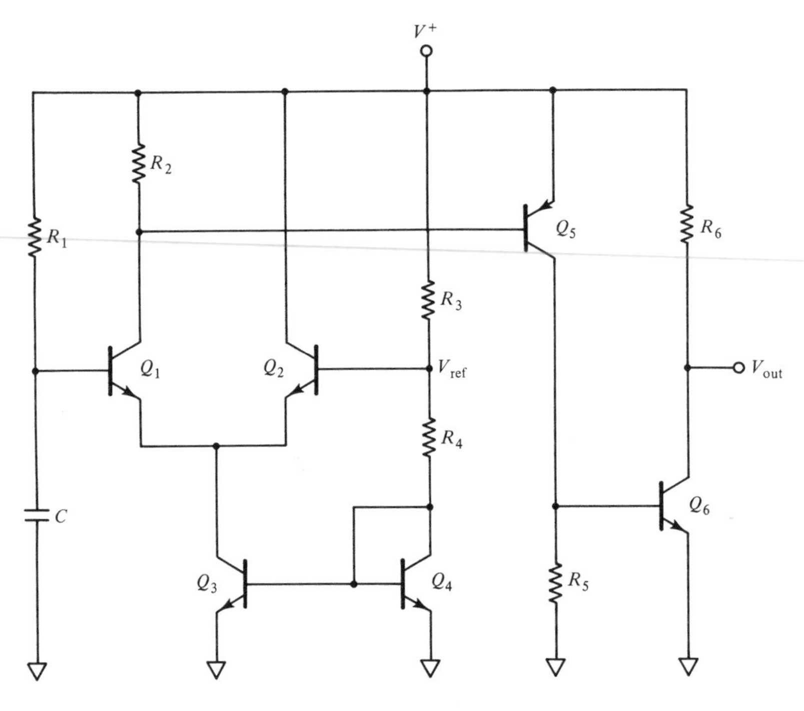

Figure 5-99 Power-up timer.

This is most likely to happen during timer reset when the capacitor is being discharged through a transistor to ground for *pnp* input comparators. Common mode range violations tend to occur on *npn* input comparators at the top of the charge cycle. The *npn* input comparators (this includes op amps and any circuit with an *npn* input pair) have another potential hazard. The maximum differential voltage that can appear across an *npn* differential pair is about 6.0 V. If one base is more than 6.0 V higher than the other base, the V_{be} of the transistor with the high base will turn on and the base-emitter junction of the transistor with the low base will begin to zener. This condition can occur if the capacitor in the timer of Figure 5-99 is being discharged by a transistor to ground and if V^+ is 12 V and V_{ref} is set from 7 to 8 V. If the comparator input common-mode voltage range or differential voltage limitation is violated, the timer performance will most likely be unpredictable.

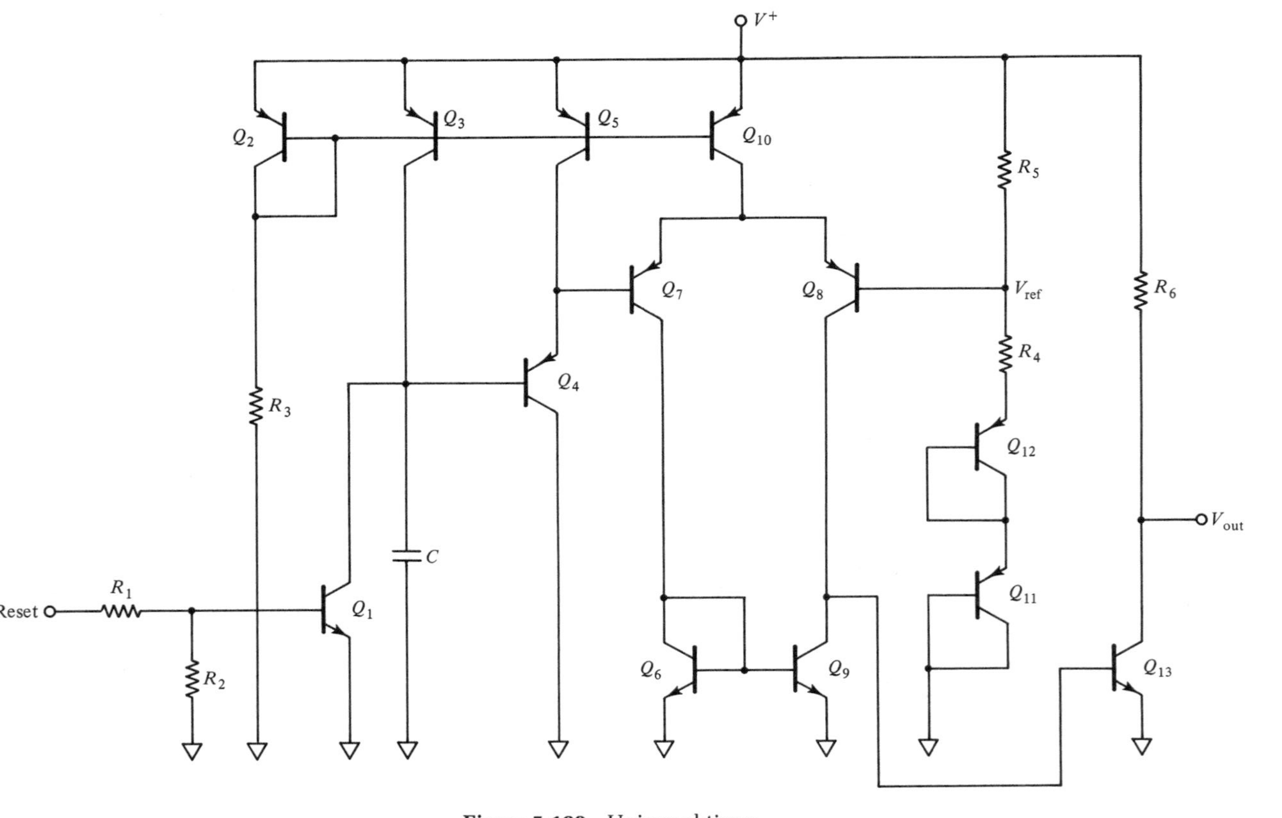

Figure 5-100 Universal timer.

5.7 Logic Circuits

Bipolar ASIC arrays are not designed to accommodate high-density logic. Frequently, however, a small amount of logic is valuable for controlling functions on an analog circuit such as gain, frequency, enable, disable, or for performing logical operations on control signals coming into the chip. Also, counters can be very useful for long-duration timers.

Logic is typically area-inefficient on an analog bipolar array unless some provision, such as a mini logic array, has been made. Logic circuitry requires many small-geometry *npn* transistors, a large number of repetitive resistor values for biasing, and lots of room for metal routing. This is contrary to the philosophy of modern analog arrays.

There are several logic types to choose from, each with its own advantages and disadvantages. ECL (emitter-coupled logic) is very fast and nonsaturating. ECL gate configurations are "linearlike" in that they make use of nonsaturating differential pairs. The disadvantages of ECL are unusual power supply voltages, high current consumption (high power dissipation), complexity, and, in most applications, inconvenient logic levels. Saturating logic families include DTL (diode-transistor logic), TTL (transistor-transistor logic), and RTL (resistor-transistor logic). DTL and TTL logic require a large number of transistors and some specialized transistor geometries. RTL represents a good compromise and is relatively simple to design.

The logic circuits presented here are RTL configurations. They are simple, easy to lay out on most arrays, easy to interface with analog circuitry, and can easily be made TTL compatible. These functions should serve most of the logic needs on an analog array. Specialized circuitry should be designed to accommodate unusual logic needs.

The logic circuits below are designed to be universal and usable on many different arrays. V^+ is intended to be +5 V for TTL compatibility. R_L should be between 2 and 10 kΩ and R_b between 1 and 5 kΩ. Other values are possible depending on the application, required speed, and the particular array technology.

NOR Gate

Figure 5-101(*a*) illustrates a typical RTL NOR gate circuit. If either Q_1 or Q_2 is turned on (*A* or *B* at a high potential or logic 1), the output

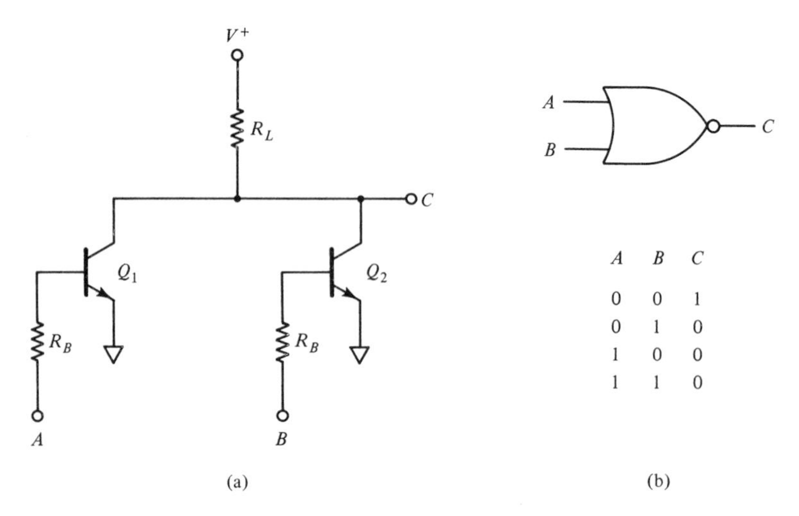

Figure 5-101 RTL NOR gate. (*a*) Circuit configuration and (*b*) logic symbol and truth table.

C will be low. The logic symbol and truth table are given in Figure 5-101(*b*).

AND Gate

Figure 5-102(*a*) illustrates a typical RTL AND gate. If Q_1 and Q_2 are turned on (*A* and *B* at a high potential or logic 1), the collector and base of Q_3 are at a potential of 2 V_{sat} ($\approx$400 mV) and Q_4's base is pulled to ground through R_L. The output *C* is high or logic 1. If either *A* or *B* is set to a logic low, the base and collector of Q_3 are pulled up to a potential of $2V_{be}$ by R_L to V^+. Q_4 then saturates, pulling *C* to a potential of V_{sat} or logic 0. The logic symbol and truth table for the circuit are given in Figure 5-102(*b*).

XOR Gate

An exclusive OR gate (XOR) is illustrated in Figure 5-103. If both inputs *A* and *B* are low (logic 0), Q_1, Q_2, Q_3, and Q_4 are cut off. R_L to V^+ and the diode-connected transistor Q_5 supply base current to Q_6, which is

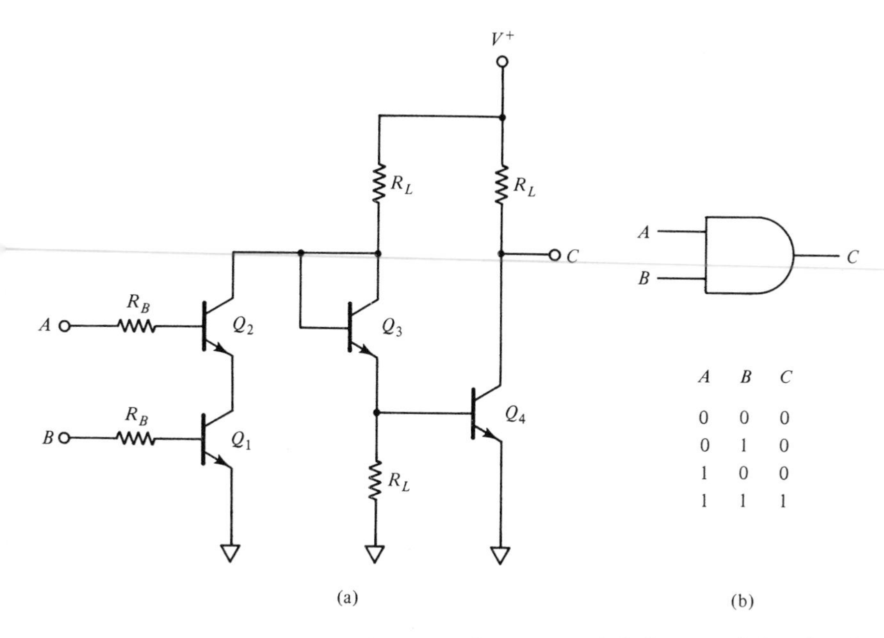

Figure 5-102 RTL AND gate. (*a*) Circuit configuration and (*b*) logic symbol and truth table.

saturated. The output C is low (at a logic 0). If input A is at a logic 1, Q_1 is saturated, cutting off Q_2 and turning on Q_3. Q_3 turning on brings the base and collector of Q_5 to a potential of $2V_{\text{sat}}$, shutting off the base drive to Q_6. Q_6 turns off, it's collector, output C, is brought to a logic 1 through R_L to V^+. If input B is at a logic 1, Q_4 is saturated, cutting off Q_3 and turning on Q_2. Q_2 turning on brings the base and collector of Q_5 to a potential of $2V_{\text{sat}}$, shutting off the base drive to Q_6. Q_6 turns off and its collector, output C, is brought to a logic 1 through R_L to V^+. If both input A and input B are at a logic 1, Q_1 and Q_4 are saturated, Q_2 and Q_3 are cut off, and R_L and Q_5 supply base current to Q_6, which is saturated, bringing output C to a logic 0.

RS Flip-Flop

An RS flip-flop is illustrated in Figure 5-104. If both set and reset are low, Q and Q' will remain in their previous state (either high or low).

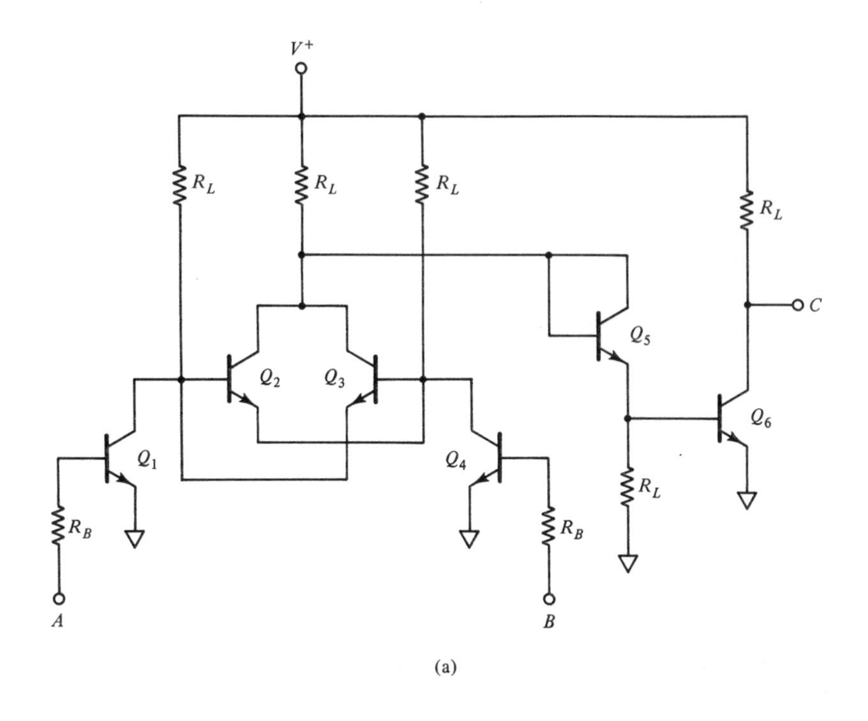

(a)

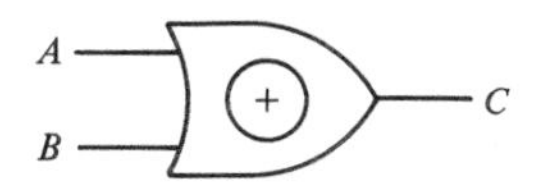

A	B	C
0	0	0
0	1	1
1	0	1
1	1	0

(b)

Figure 5-103 RTL XOR gate. (*a*) Circuit configuration and (*b*) logic symbol and truth table.

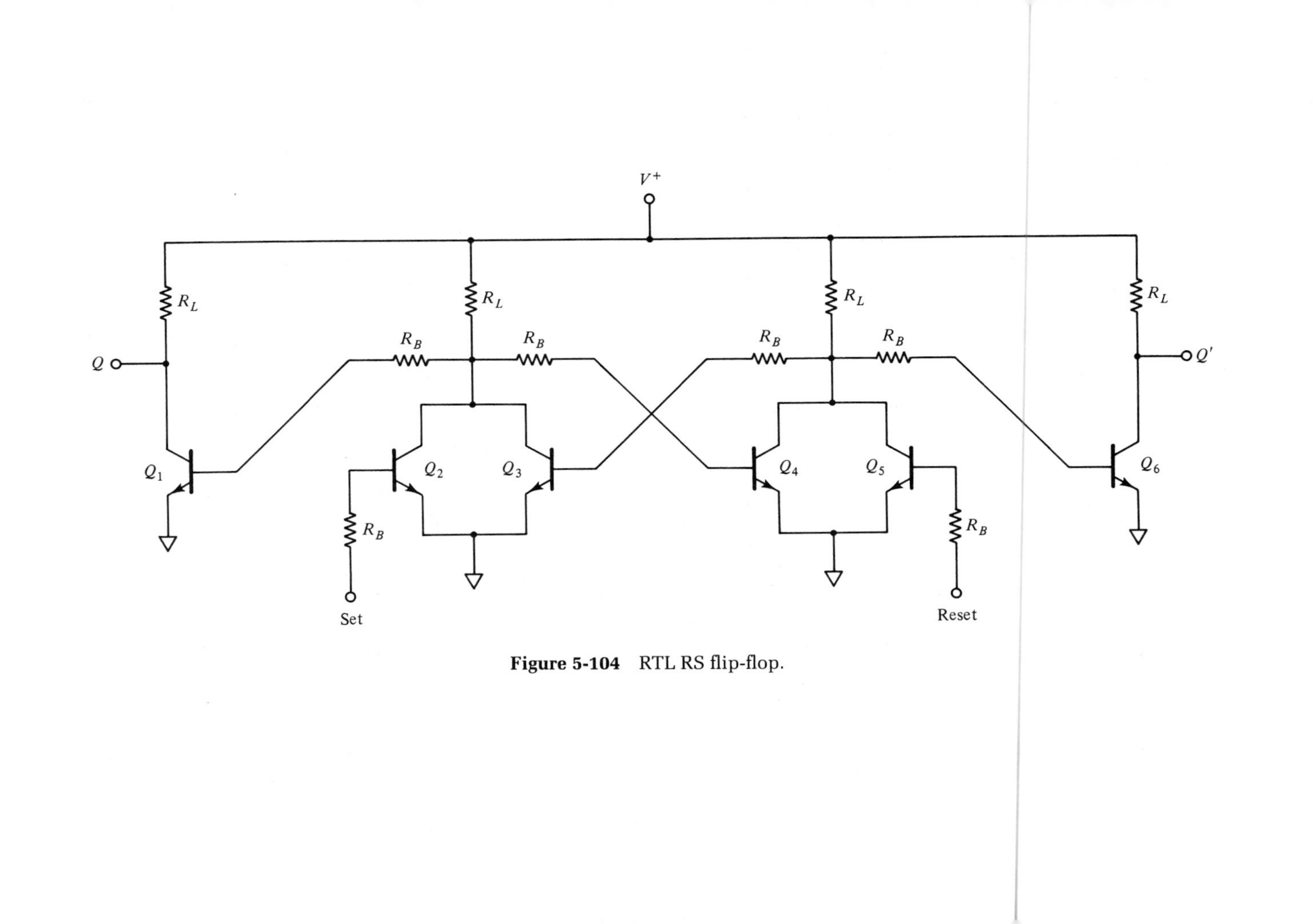

Figure 5-104 RTL RS flip-flop.

If set goes high and reset remains low, Q_2 saturates and turns off Q_1 and Q_4. Output Q is set to a logic high and output Q′ is set to a logic low. If set goes low, the states of Q_3 and Q_4 reinforce this condition and no change occurs. If reset goes high with set held low, Q_5 saturates, turning off Q_3 and Q_6. Q′ goes to a logic 1 through R_L to V^+. Q_1 saturates and Q goes to a logic low. Q_3 and Q_4 reinforce this condition. If both set and reset are brought high and then low simultaneously, the states of Q and Q′ will be indeterminate.

Clocked Latch

The clocked latch shown in Figure 5-105 operates similarly to the flip-flop in Figure 5-104, except that the outside transistors in the emitter-coupled pair are enabled only when the clock is high. The data provided to Q_3 is inverted from that provided to Q_6. This creates a condition similar to a set in Figure 5-104 when the clock transistor Q_2 is saturated (clock at logic 1). When the clock goes low, Q_4 and Q_5 reinforce the last data condition.

For more information on logic circuits see References 38 and 39.

5.8 Circuit Configurations

It has been said that synthesis is analysis plus experience. Many engineers setting out on an ASIC design may not have experience with some useful circuit concepts. The purpose of this section is to provide a variety of circuit concepts that will help stimulate some creative ideas. It is very difficult, unless you are an experienced IC designer, to sit down and start drawing circuit schematics. There are many times, however, when this depth of knowledge is very helpful. Having a good preliminary estimate of the number and type of components will allow you to choose, or eliminate, potential arrays for your proposed ASIC. Knowledge of the component density of various implementations of a circuit will help you to make decisions regarding the optimum implementation, taking into account circuit complexity and performance. Finally, when actual circuit design begins, having a detailed, component-level schematic will give you a starting point from which to begin design. The initial schematic may well be significantly modified by the time design work is complete, but it is much better to start with something

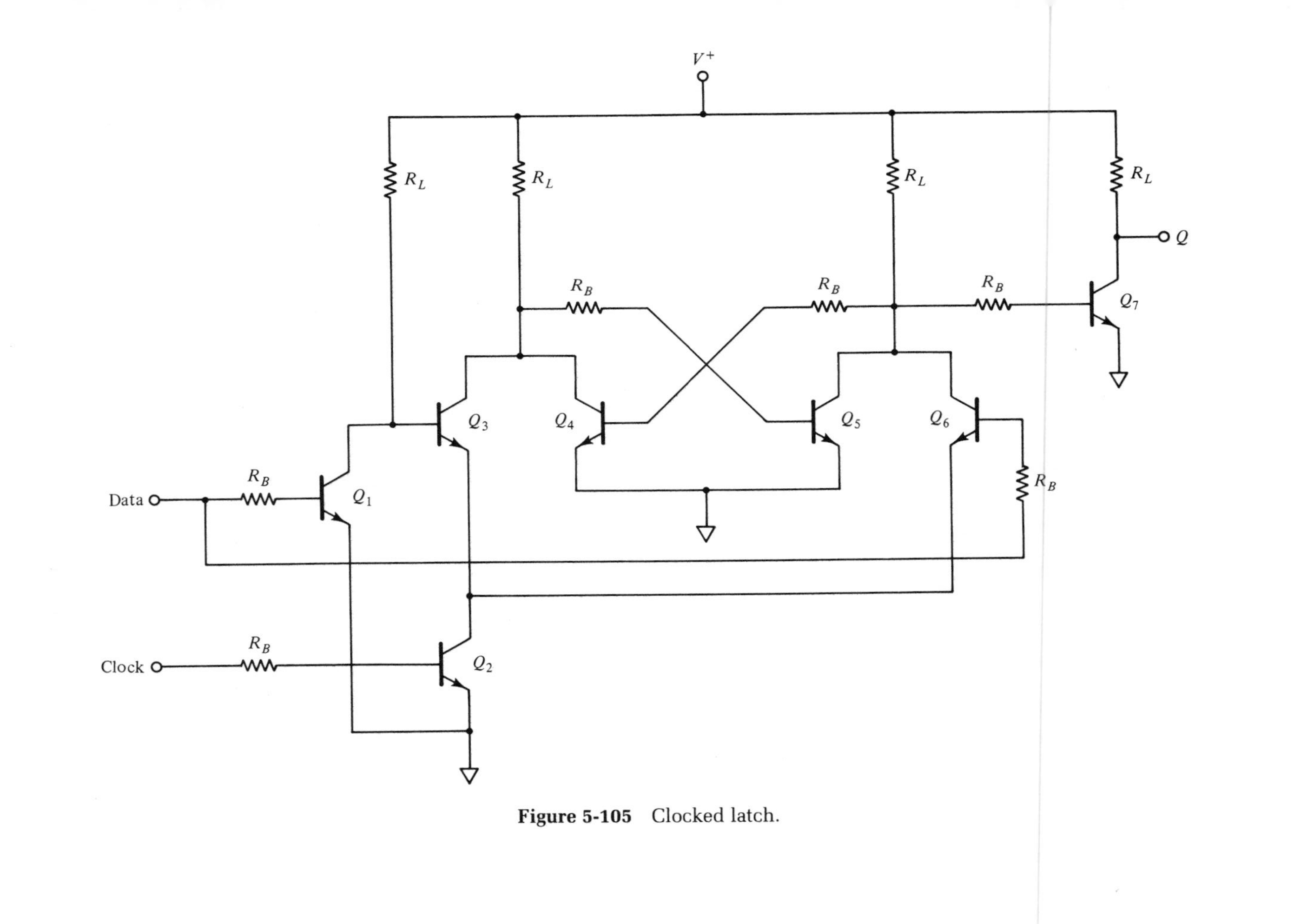

Figure 5-105 Clocked latch.

and modify it to suit your needs rather than to start with a blank page. This section is not meant to be an all-inclusive or even a significant library of circuits. See Reference 39 for a large number of detailed transistor-level schematics of integrable circuits.

The functional block schematics and application schematics using these functional blocks are meant to provide concepts for possible implementations of various circuits. Component values have been omitted due to the different values available from vendor to vendor and the wide variation in array performance. An amplifier may work fine on a relatively tame 20-V, 350-MHz ASIC array but could oscillate furiously on a 12-V, 1-GHz array. Many times a change in bias point or bias current, or the addition of a compensation capacitor, will adapt an amplifier to various arrays. The basic circuit configuration, however, will remain the same.

The circuits shown in the collection of op amp circuits do not necessarily require a full blown "op amp" to function. The op amp symbol should be viewed as a gain block. These circuits will work just fine with an op amp (depending on performance requirements). Some

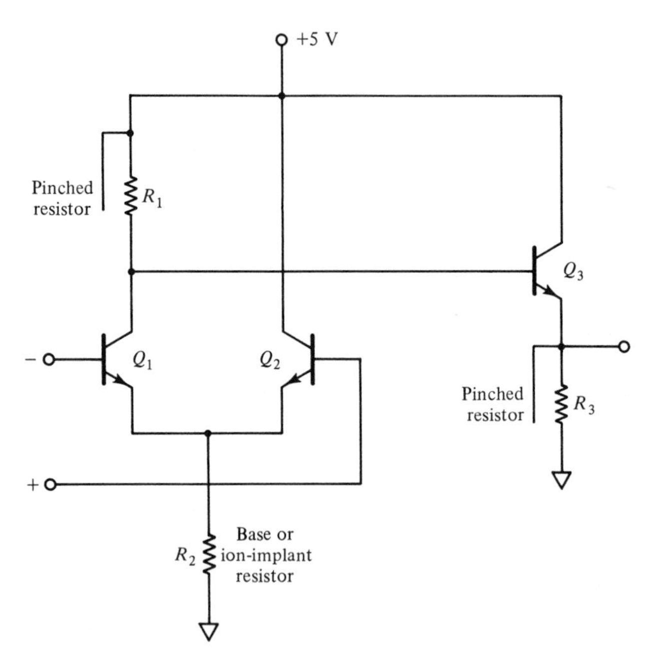

Figure 5-106 Simple differential input amplifier.

op amp configurations, however, may be a significant overkill in terms of the number of components needed to perform that particular function. For example, if an amplifier's output does not leave the IC, short-circuit current limit is probably not indicated. In other cases, an "op amp" will not have the high-frequency response necessary to perform a given function in a particular application. In most semicustom designs, component efficiency is very important. The fewer components required to perform a function, the more circuitry can be integrated or the smaller (less expensive) the array that will be required to integrate the circuitry. Throughout the design process trade-offs such as design time versus manufacturing cost or overdesign versus marginality must be made continuously. Standard-product IC data books and IC schematics are a good source of circuit design ideas. With these caveats in mind, Figures 5-106 through 5-130 illustrate some useful circuit configurations for arrays and standard cells.

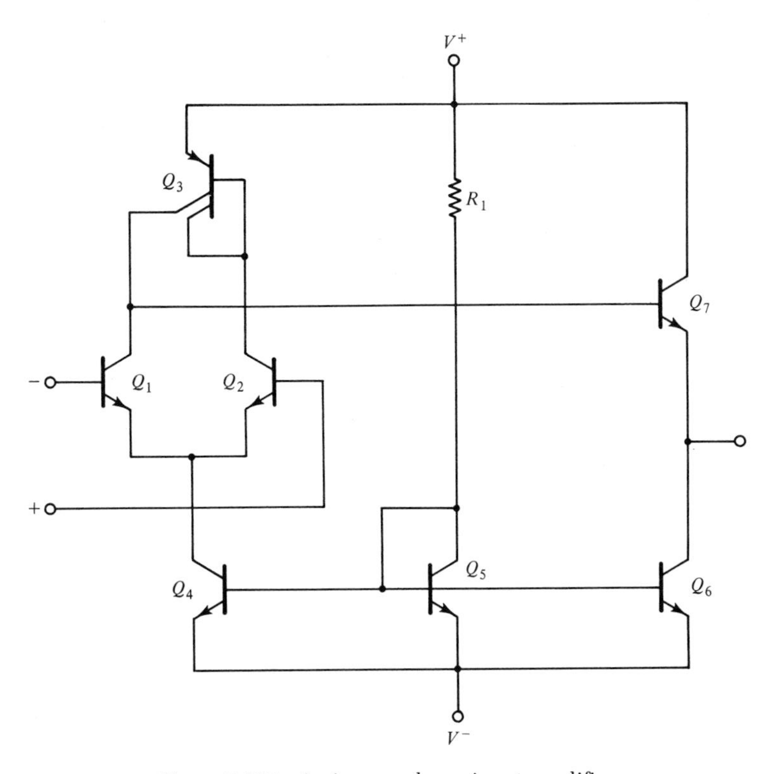

Figure 5-107 An improved *npn* input amplifier.

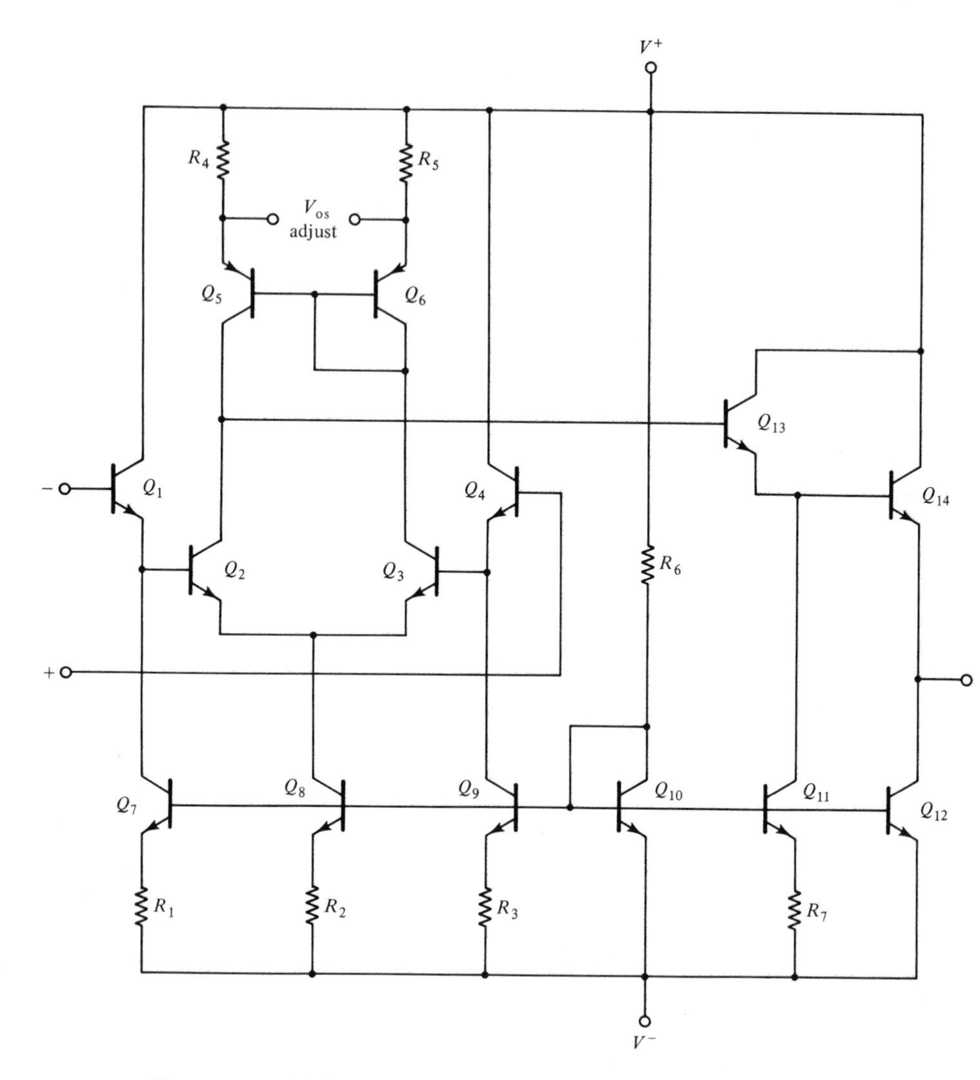

Figure 5-108 High-input impedance *npn* input amplifier with V_{os} adjust.

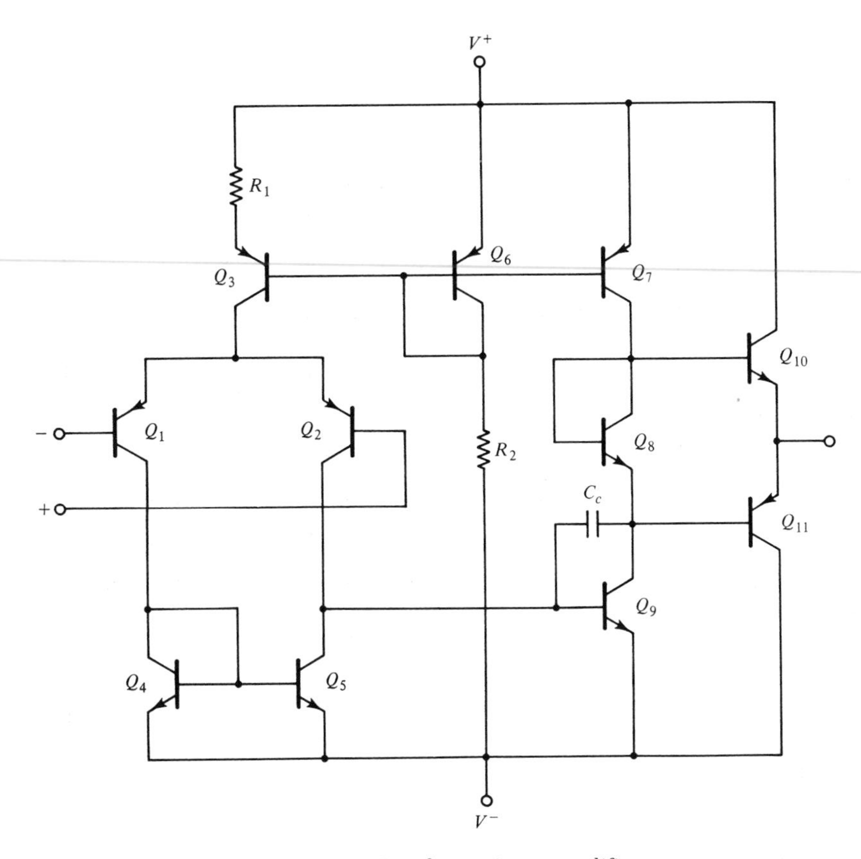

Figure 5-109 Simple *pnp* input amplifier.

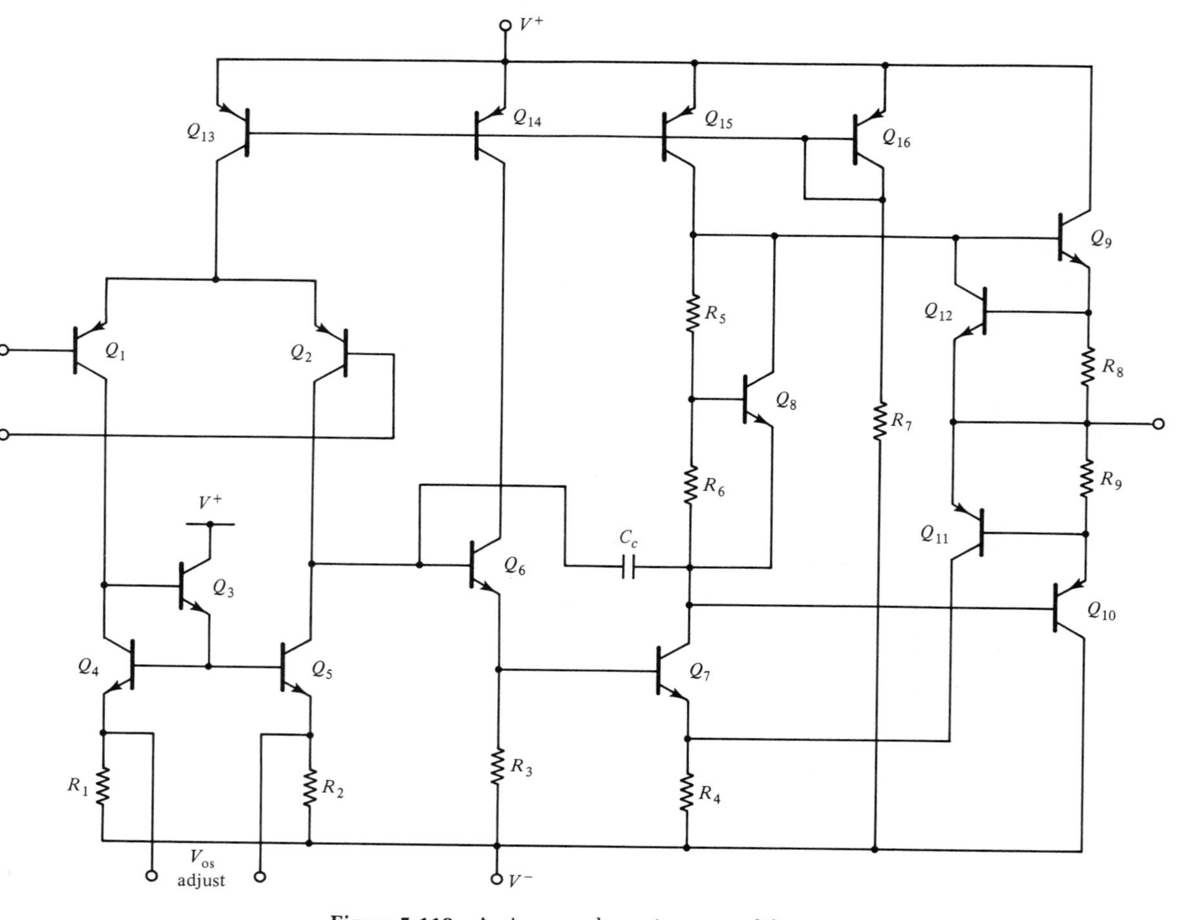

Figure 5-110 An improved *pnp* input amplifier.

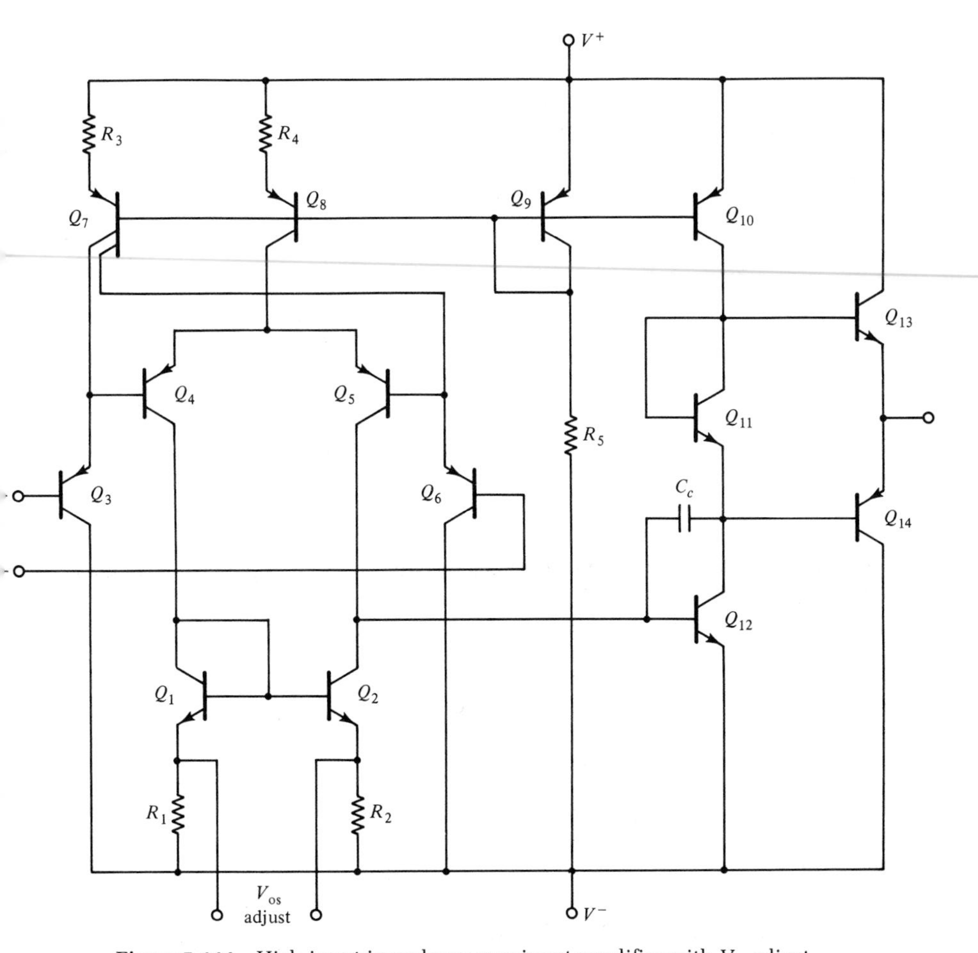

Figure 5-111 High-input impedance *pnp* input amplifier with V_{os} adjust.

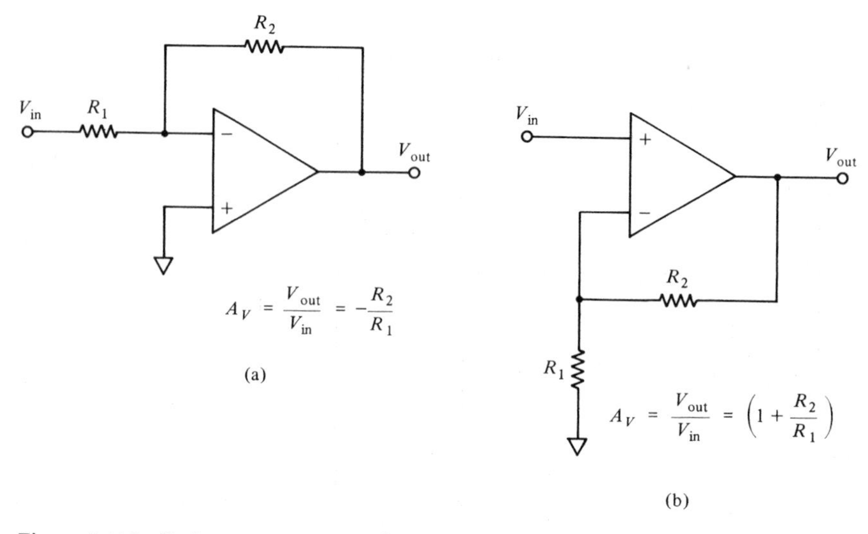

Figure 5-112 Basic op amp gain configurations. (*a*) Inverting amplifier and (*b*) noninverting amplifier.

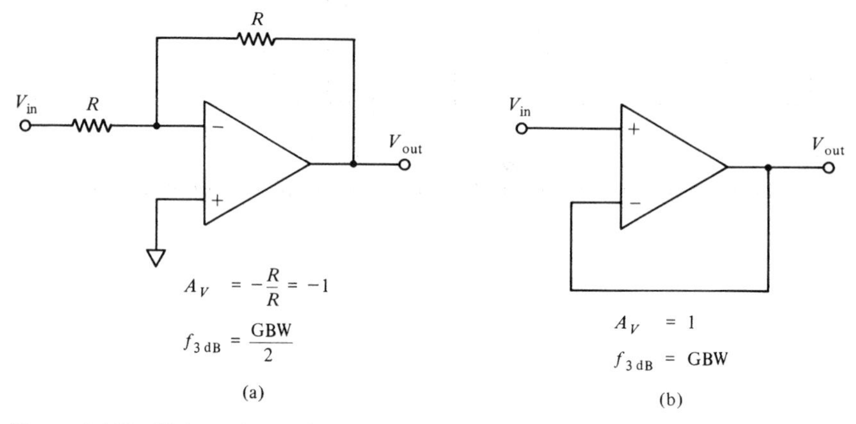

Figure 5-113 Unity gain configurations. (*a*) Unity gain inverter and (*b*) unity gain follower.

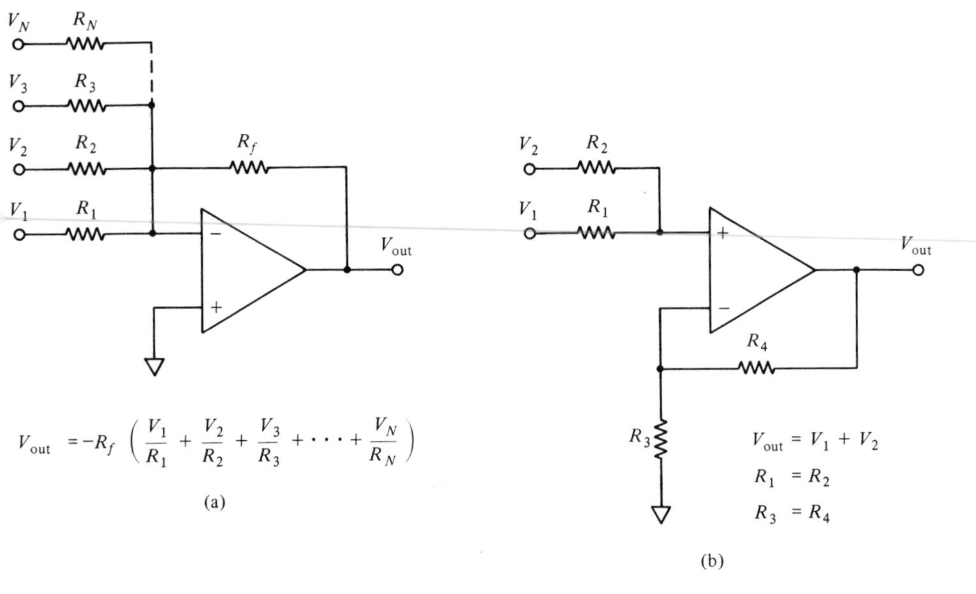

Figure 5-114 Summing amplifiers. (*a*) Inverting summing amplifier and (*b*) noninverting summing amplifier.

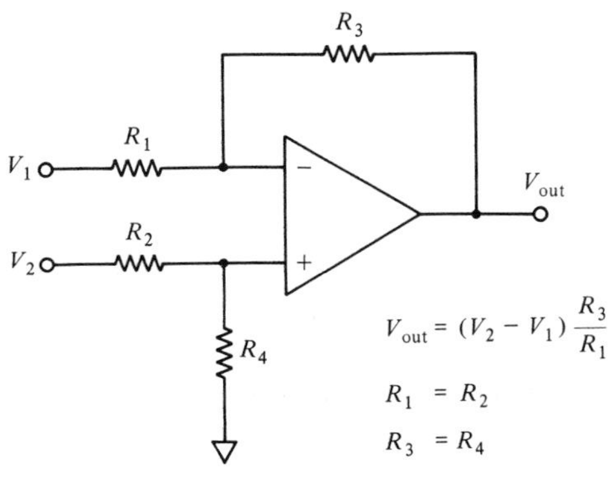

Figure 5-115 Differential amplifier.

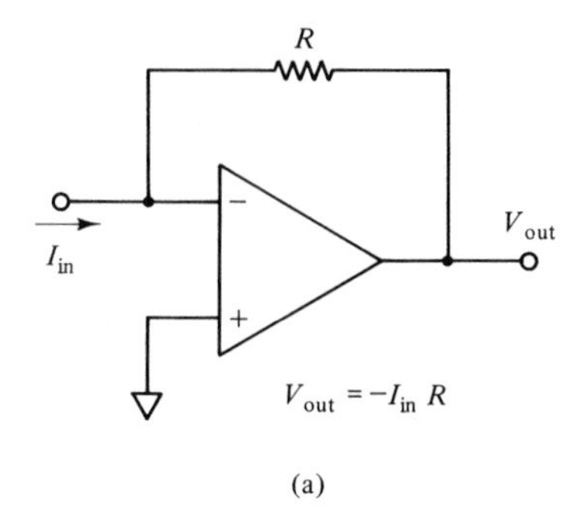

(a)

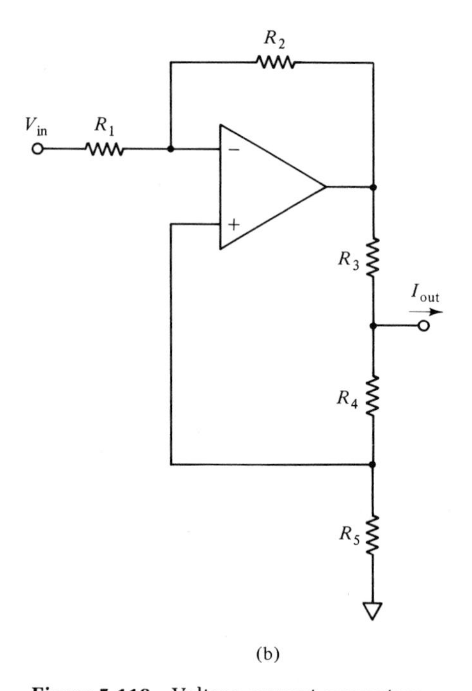

(b)

Figure 5-116 Voltage-current converters.

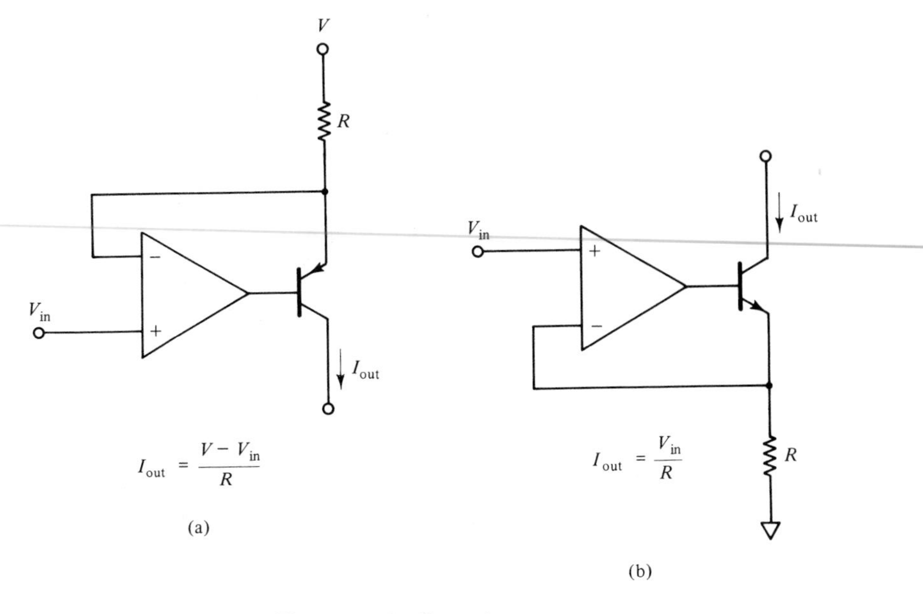

Figure 5-117 Current sources.

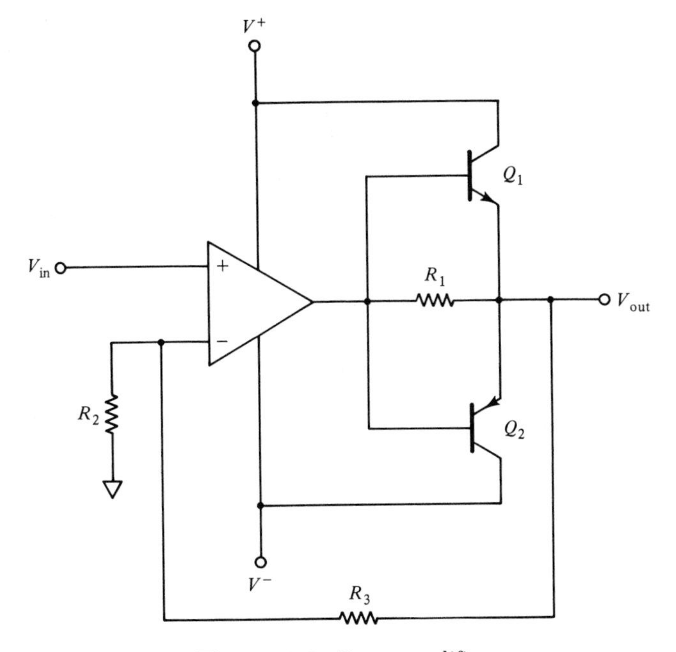

Figure 5-118 Power amplifier.

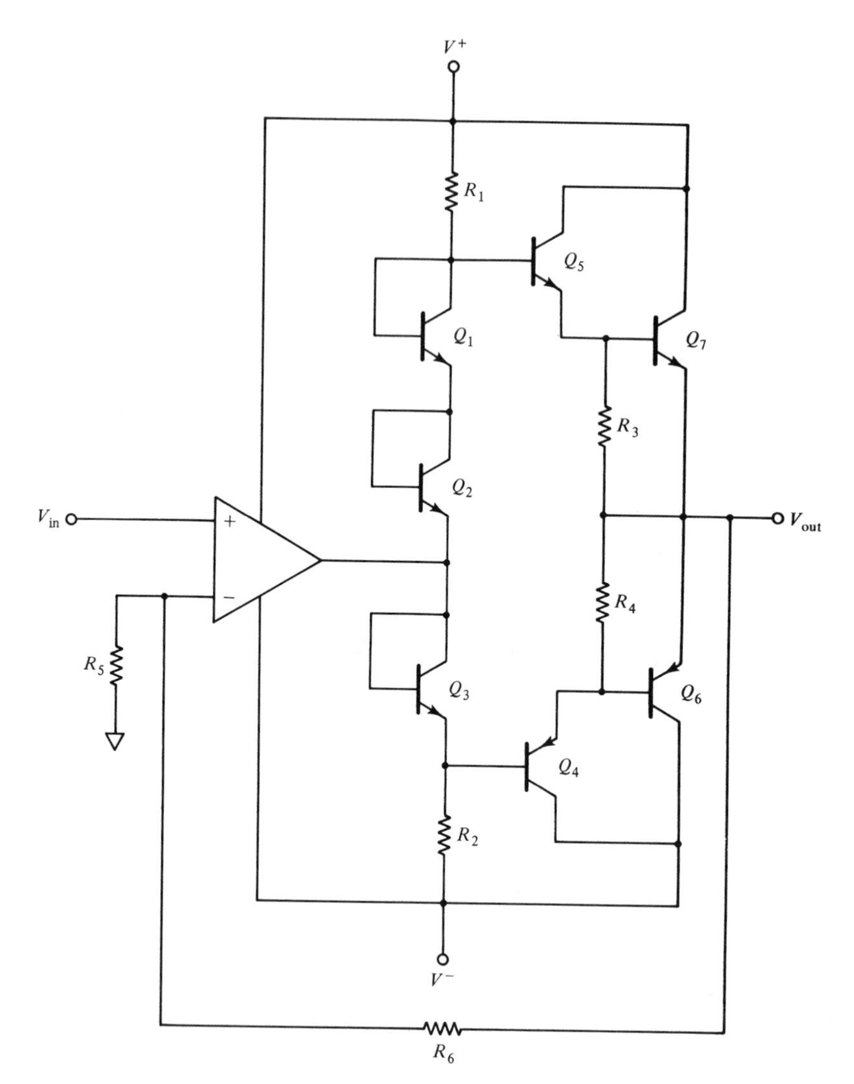

Figure 5-119 Improved power amplifier.

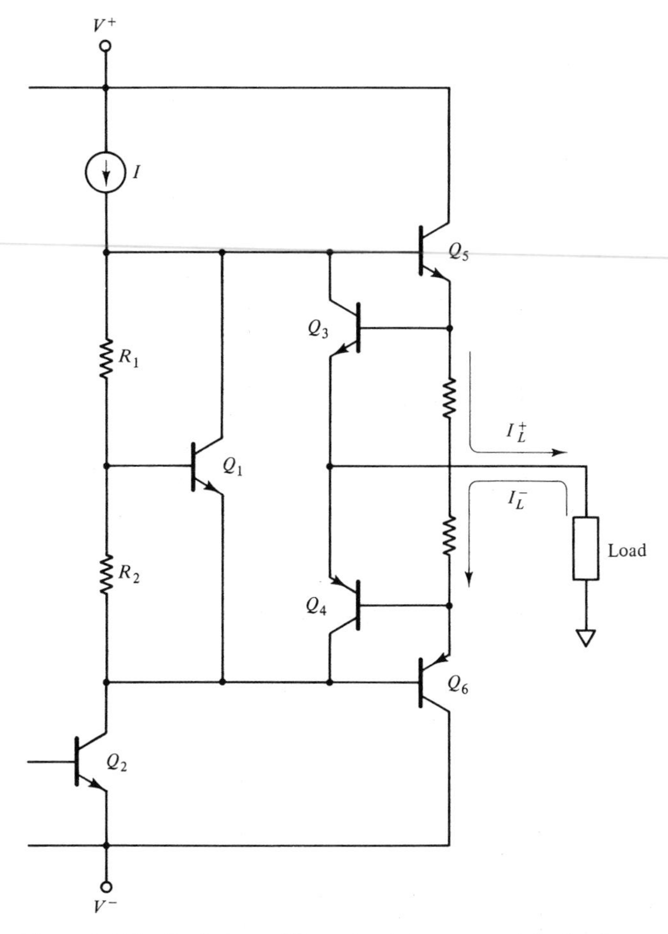

Figure 5-120 Typical amplifier output-stage current-limit scheme.

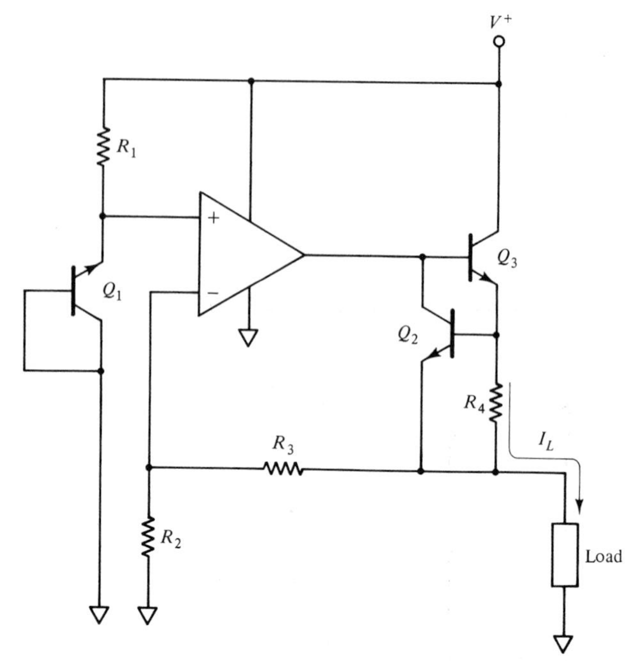

Figure 5-121 Voltage regulator with current limit.

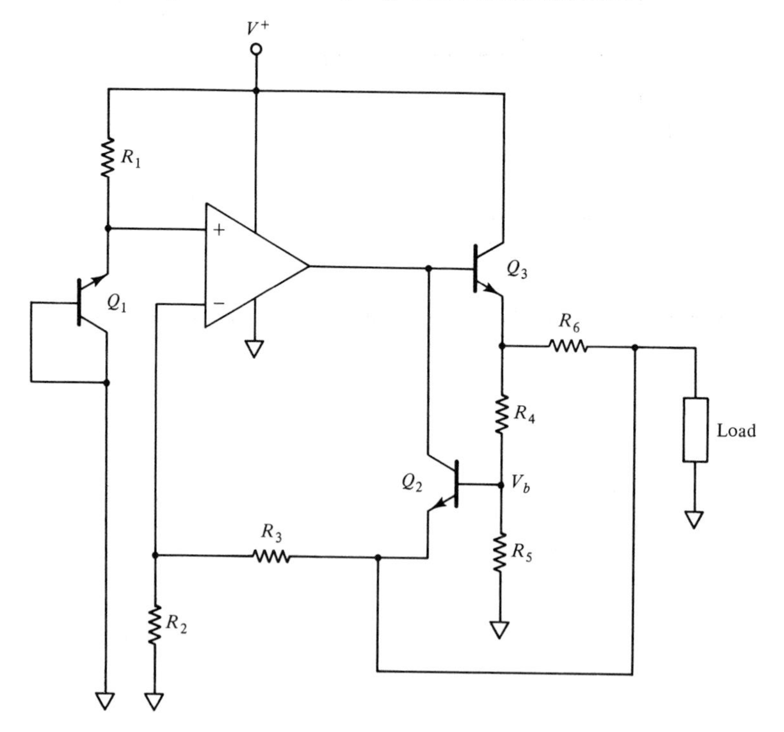

Figure 5-122 Voltage regulator with foldback current limit.

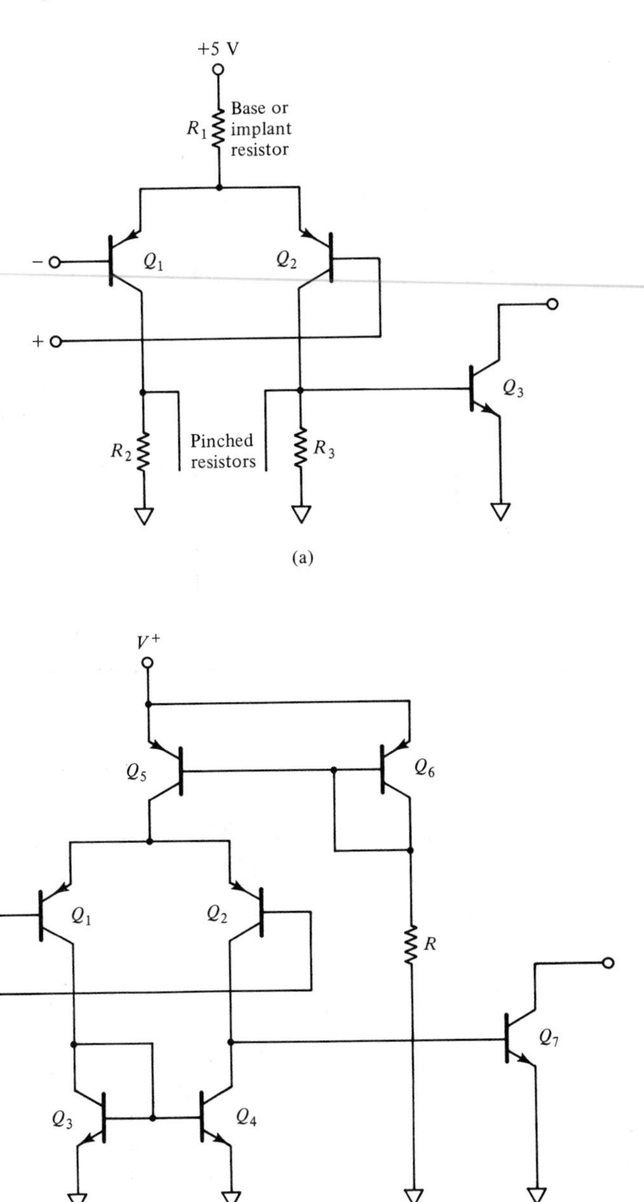

Figure 5-123 Simple comparator circuits. (*a*) Positively biased and (*b*) current-source biased.

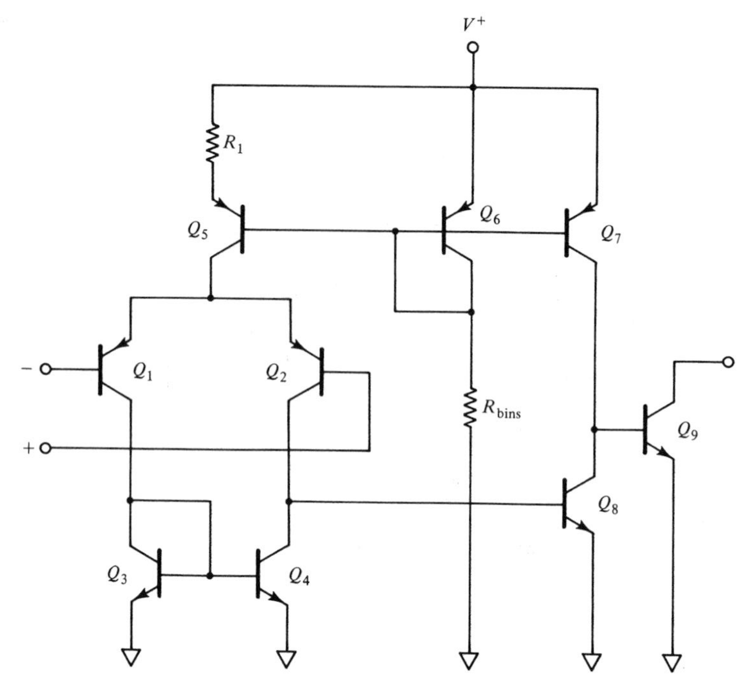

Figure 5-124 Simple comparator with improved output drive.

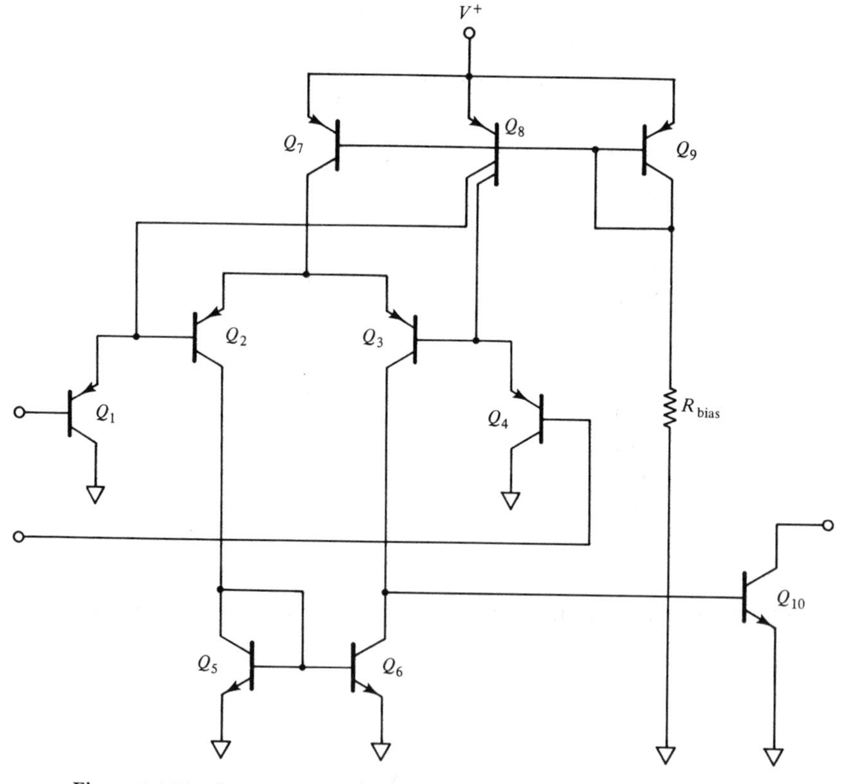

Figure 5-125 Comparator with common-mode voltage, including ground.

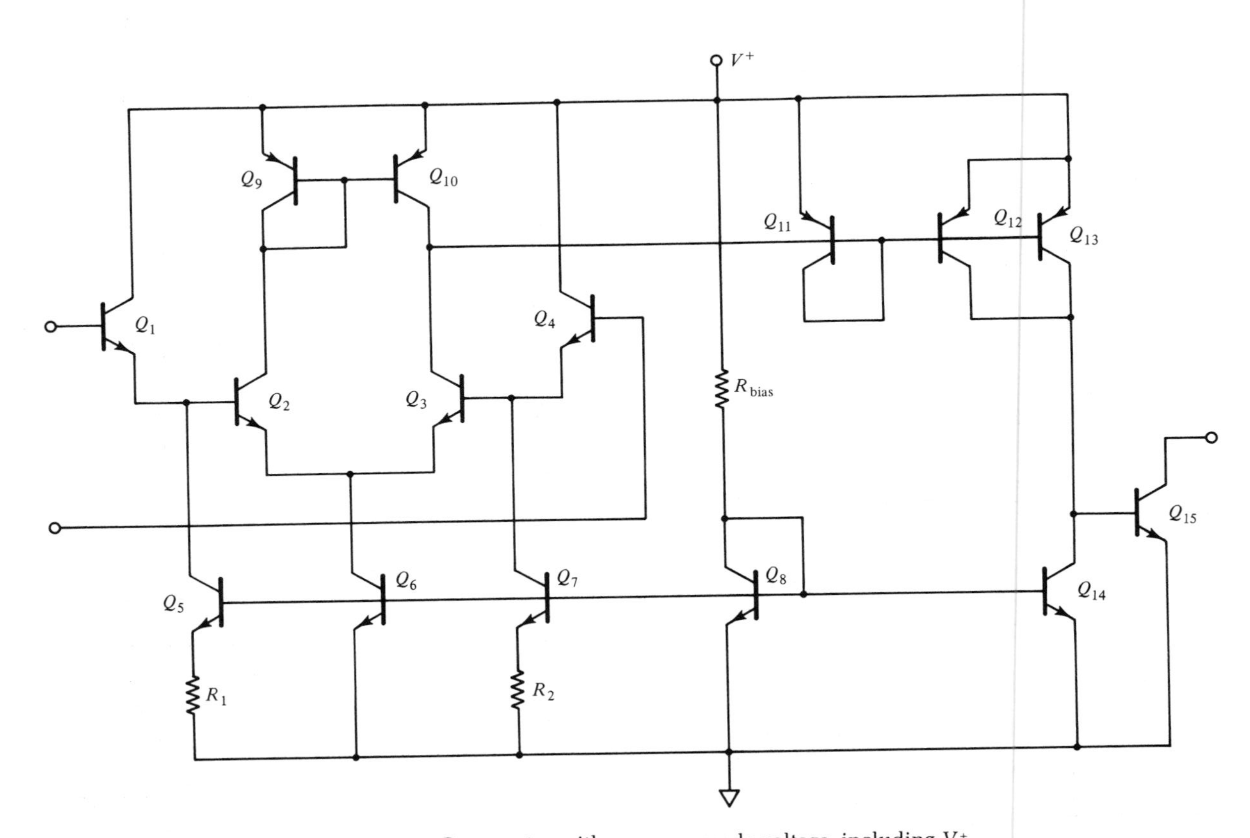

Figure 5-126 Comparator with common-mode voltage, including V^+.

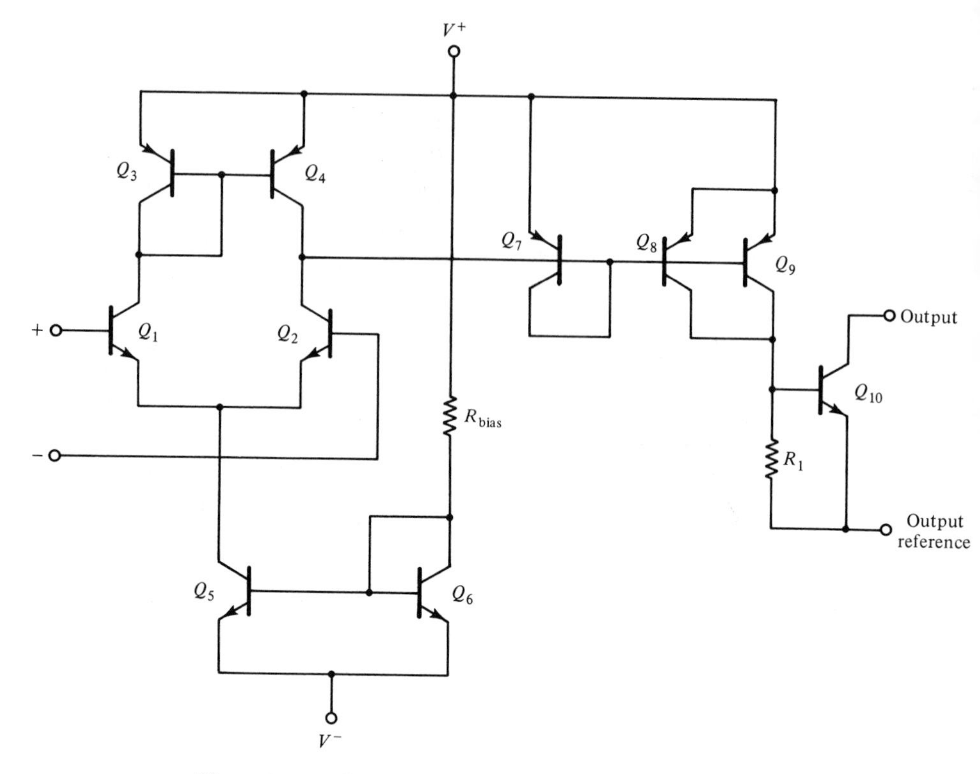

Figure 5-127 Comparator with independent output reference.

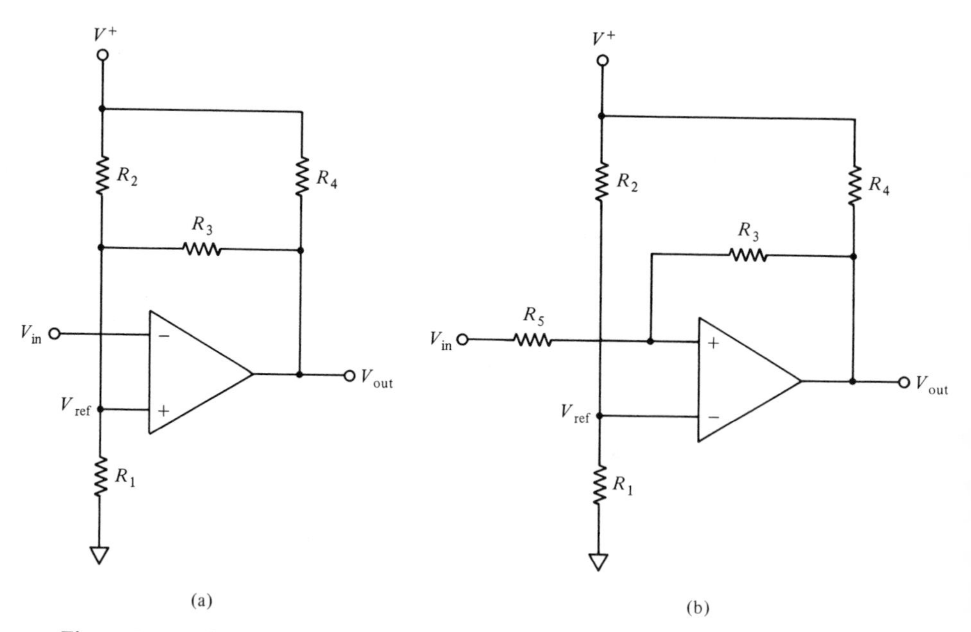

Figure 5-128 Comparators with hysteresis. (*a*) Inverting configuration and (*b*) noninverting configuration.

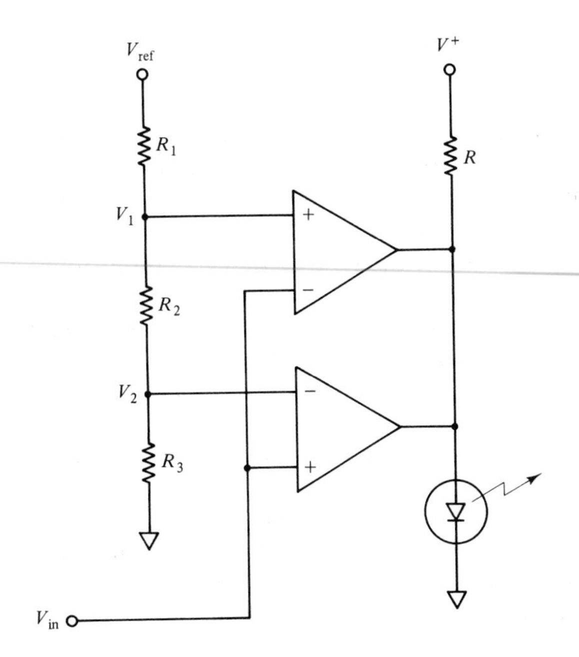

Figure 5-129 Limit detector.

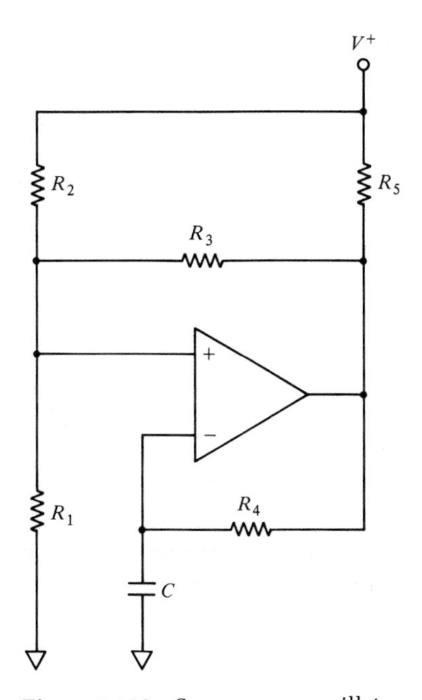

Figure 5-130 Square-wave oscillator.

5.9 Worst-Case Considerations

During the creative design process and at the completion of detailed circuit design, the performance of the circuit under "worst-case" conditions must be reviewed. The problem is identifying and understanding just what the worst-case conditions are. They may include high temperatures or low temperatures or even room temperature. Worst case may be high *npn* beta or low *npn* beta and high resistor values, or the variation of device characteristics after prolonged exposure to radiation. What is worst case for an IC as a whole is not obvious. Parametric variations that affect one subcircuit on the die may have little or no effect on another.

Take a top-down approach from the system level to the circuit level to identify the true worst-case conditions. Never forget that the ASIC is an integral part of a system and the system performance requirements will drive the necessary IC performance. The following steps will help drive out the real worse-case conditions and will identify potential quality, reliability, and manufacturability problems.

- **Define system-level performance requirements.** This data will have been prepared during the initial system partitioning and IC definition process.
- **Define all operating conditions.** The operating conditions include the environmental conditions such as temperature, shock and vibration, radiation, external load, and impedance variations and power supply variations.
- **Define all parametric variations of signals and components external to the IC.** Components and signals external to the IC will vary with the environmental conditions, with age, and with unit-to-unit tolerances.
- **Define the range of conditions that will appear at each IC pin.** Develop a table that defines the range of signal levels and/or impedances for each IC pin for each environmental and component parametric variation, then combine these variations to estimate the worst-case conditions external to the chip at each pin.
- **Partition the IC component level circuit into subcircuits.** This was done during the initial IC definition phase.
- **Define all parametric variations expected on the IC by component for each subcircuit.** These variations include beta, resistor value, and f_t variations due to process tolerances, radiation effects, and temperature changes.

• **Given the operating conditions and parametric variations, evaluate the performance of each subcircuit.** SPICE analysis is a useful tool for performing multiple-circuit simulations with changing component values and signal levels.

• **Evaluate the performance of all subcircuits with respect to the system.** SPICE analysis or breadboarding can be used to closely emulate the actual device performance.

• **Identify areas of marginal performance.**

• **Make any necessary design changes.**

• **Carefully document this analysis and any design modifications.** Performing a rigorous worst-case analysis will identify design marginalities during the design phase. It will also serve to validate the design approach by thoroughly reviewing the anticipated performance.

Worst-case analysis is the most valuable if it is conducted by an engineer who has not contributed to the design. An independent analysis will serve as a validation of the entire design concept.

References

1. Gray, P. R., and Meyer, R. G., *Analysis and Design of Analog Integrated Circuits*. Wiley & Sons, New York, 1977; p. 197.
2. Grebene, A. B., *Bipolar and MOS Analog Integrated Circuit Design*. Wiley & Sons, New York, 1984; p. 175.
3. Gray and Meyer, (1977), p. 201.
4. Gray and Meyer, (1977), pp. 208–210.
5. Grebene, (1984), pp. 180–181.
6. Gray and Meyer, (1977), pp. 199–201.
7. Grebene, (1984), pp. 181–183.
8. Gray and Meyer, (1977), pp. 252–253.
9. Grebene, (1984), pp. 195–196.
10. Gray and Meyer, (1977), pp. 254–261.
11. Grebene, (1984), pp. 206–212.
12. Comer, D. J., *Introduction to Semiconductor Circuit Design*. Addison-Wesley, Reading, Mass., 1968; Chapter 5.
13. Comer, (1968), Chapter 7.
14. Solomon, J. E., The monolithic op amp: A tutorial study, *IEEE Journal of Solid-State Circuits*, **SC-9**, (6), December 1974.
15. Grebene, (1984), Chapter 7.
16. Gray and Meyer, (1977), Chapter 6.
17. Boyle, G. R., Cohn, B. M., Pederson, D. O., and Solomon, J. E., Macromodeling of integrated circuit operational amplifiers, *IEEE Journal of Solid-State Circuits*, **SC-9** (6), December 1974.
18. Smathers, R. T., Frederiksen, T. M., and Howard, W. M., *LM139/LM239/LM339 A*

Quad of Independently Functioning Comparators. National Semiconductor, Santa Clara, CA, 1973, AN-74.
19. Comer, (1968), pp. 213–215.
20. Bohn, D., *Audio Handbook*, National Semiconductor, Santa Clara, CA, 1976; pp. 6–14.
21. Comer, (1968), pp. 366–370.
22. Comer, (1968), p. 370.
23. Frerking, M. E., *Crystal Oscillator Design and Temperature Compensation*. Van Nostrand Reinhold, New York, 1978; Chapter 5.
24. Orr, W. I., *Radio Handbook*, 20th ed., Howard W. Sams, Indianapolis, 1975; pp. 11.10–11.14.
25. Grebene, (1984), pp. 546–547.
26. Strauss, L., *Wave Generation and Shaping*. McGraw-Hill, New York, 1970; pp. 669–671.
27. Strauss, (1970), pp. 666–668, 671.
28. Strauss, (1970), pp. 675, 690.
29. Grebene, (1984), p. 547.
30. Grebene, (1984), Section 11.3.
31. Grebene, (1984), Chapter 9.
32. Gray and Meyer, (1977), Section 10.3.
33. Gilbert, B., High-accuracy vector sum and vector difference circuits, *Electronics Letters*, **12**, 293, May 1976.
34. Ashok, S., Translinear root difference-of-squares circuit. *Electronics Letters*, **12**, 194, April 1976.
35. Gilbert, B., Translinear circuits: A proposed classification, *Electronics Letters*, **11**, (1), 14, January 1975.
36. Gilbert, B. General technique for *n*-dimentional vector summation of bipolar signals, *Electronics Letters*, **12**, (19), 504, September 1976.
37. Grebene, (1984), pp. 314–315.
38. Comer, (1968), Chapter 9.
39. Staff, *201 Analog IC Designs*, Interdesign, Scotts Valley, CA, 1980; pp. 40–44.

CHAPTER 6

Integrated Circuit Layout Considerations

6.0 Introduction

Integrated circuit layout is part of the integrated circuit design. All of the components on an IC are made of the same material and are thermally and electrically connected through the substrate. Heat generated on the chip is distributed throughout the chip. It is conducted through the substrate to the package, through the bonding wires to the pins, and dissipated by the package into the ambient. Each component on the chip is subjected to the temperature variations of the other components. The matching of components is greatly affected by their relative positions and orientations even though they are all made on the same material concurrently. Thermal gradients from components dissipating relatively large amounts of power to relatively cooler areas on the chip can induce signals in the form of thermally induced offsets. These "signals" can change with variations in power dissipation caused by external load changes or common-mode voltage changes. Metal interconnections on the chip, although short by PC board standards, are very thin and narrow and exhibit a significant ohmic resistance. Voltage drops on the metal can cause unwanted signal injection and feedback. Similar problems can be caused by current being injected into the substrate. Circuit performance is impacted to a far greater extent by circuit layout on an IC than on a printed circuit board.

Each vendor's product and fabrication process is different. Even if these differences are slight, it is not unreasonable to expect differences in layout rules. An ASIC user planning to do a layout in-house must get up-to-date layout data from the vendor (or vendors) whose products are intended for use before any layout work is started. Many vendors offer layout packages that include instruction manuals, guidelines, software, or hard copy scale drawings of the chip on which to indicate metal routing. If you are considering two sources (each offering equivalent arrays) it would be wise to do the layout using the most conservative layout rules between the two vendors. Then a single metal routing would be manufacturable by both vendors. This could save time if you wish to buy product from the alternate vendor and would avoid any uncertainties regarding circuit performance due to a layout difference.

6.1 Choosing the Array

Most custom analog array vendors offer a variety of arrays in various sizes from very small to large. Some offer almost exact equivalents of those offered by other vendors. The different size arrays vary not only in the number of components but type of components, number of bonding pads, arrangement or grouping of components, and process. Generally speaking, the smaller arrays are less expensive. Choosing the "best" array for your particular application requires several different, and at times conflicting, considerations.

Process

The array you choose must be made with a semiconductor process that meets basic circuit functionality requirements. The process must have an adequate breakdown voltage. Most vendors offer arrays with maximum operating voltages of 12 V, 20 V, and 36 V. The absolute maximum voltages are generally higher but a safety margin is desirable. The speed of the process must meet your circuit performance needs. Generally, the lower voltage processes have the highest speed/bandwidth performance. For example, a typical 12-V process may offer an *npn* f_t of 1 GHz, while a typical 20-V process may have an *npn* f_t of 400 MHz and that for a 36-V process may only be 350 MHz. A conflict can occur when you need to handle supply voltages in the 30-V (±15 V) range but also need 1 GHz *npn* f_t. In these cases the circuit partitioning on the system level needs to be revisited. Consideration may be given to adding external voltage regulators, level shifters, or repartitioning the circuitry such that the high-frequency or high-voltage portion of the circuitry is handled off-chip.

Second Source

If possible, you will want to consider an array with equivalents offered by multiple vendors. While these "equivalent" arrays are not exactly the same (there are always some dimensional differences), the component placement and device geometries are very similar and the same metal routing can generally be used on any of them. New tooling will have to be generated for each vendor but this is a minor effort once the routing is defined. Using a multiply sourced base array will allow your custom IC to be quickly and inexpensively supplied by a second vendor should problems occur with your first vendor.

A different approach to an alternate source is to identify an array manufactured by a different vendor that is not physically the same or even made by the same process but could be used to integrate your circuit. Several issues must be carefully considered in this case. The layout and relative component placements would be different on the products from the two sources. There may, as a result, be differences in the matching and thermal tracking of the circuitry. Differences in "standard"-array resistor values may cause bias currents and, therefore, performance to vary from one product to another. Differences in transistor geometries or processes may affect the high-frequency performance of one or the other array. This could be good if high-frequency performance was improved by the alternate source or bad if unwanted oscillations or overshoots appear. External component value changes on compensation capacitors or bias-setting resistors may be required to accommodate the different devices. Verify that both the prime and the alternate source array will fit in in the desired package. An alternate die in an incompatible package is not very helpful. In any case, alternate source devices should be marked such that they are readily distinguishable. This will save countless hours of time troubleshooting should problems occur.

Number and Type of Components

The array you select will have to include the type of components or cells in sufficient quantity to integrate your circuitry. Pay particular attention to high-current transistors, bonding pads, and other specialized geometry devices or cells. You will generally be able to use all of these specialized devices. The general-purpose transistors and resistors will not all be usable due to metal routing constraints. Some components, such as multiple-collector-contact *npn* transistors or resistors, may be sacrificed for use as cross-unders. This is more of a concern on arrays with one metal layer. In this case, diffusions must be used to allow metal traces to cross without connecting. Some arrays have special resistors with very low values for this purpose. Others require the use of regular components.

Arrangement or Grouping of Components

Different arrays have different component groupings. Choose an array with a component grouping that lends itself to your particular circuit topology whenever possible. Some older arrays have an almost random arrangement of components. These were originally designed for the integration of resistively biased discrete transistor circuitry. Other arrays have a somewhat regular grouping of components. Modern analog arrays are organized in a cellular format oriented toward building block-type designs. Modern arrays are designed to accommodate current source biasing, differential circuitry, and standard building blocks such as op amps and comparators. Some vendors have predesigned and characterized macrocells that overlay on their arrays. Using these can greatly simplify the design and layout of a custom IC and significantly reduce the design risk.

The relationship between the circuitry to be integrated and the component grouping on the array will determine both the layout efficiency and, to some extent, the achievable circuit performance. For example, attempting to lay out an IC with several op amp–like circuits on an older type array with randomly placed components will result in several significant compromises. The layout efficiency will be poor, thus requiring a larger (more expensive) array. The circuitry will not be tightly laid out. An op amp may be sprawled over a quarter of the chip. The available components may not be oriented correctly, thus inducing mismatches. Circuitry that is spread out over a large portion of the chip is more susceptible to component variations and thermally induced offsets than compactly laid out circuitry. Long metal traces with their attendant parasitic resistance could be required to distribute power, ground, bias, and signals. Layout efficiency in this case would be on the order of 60 to 70%.

On a modern cellular array, the same circuit could achieve as high as 90% layout efficiency. The use of macrocells would make the layout effort a much higher level task since macrocell blocks such as op amps would be used like components. These cells are designed and laid out by the vendor to use the available components efficiently. They are designed by experts to provide good matching and thermal performance. The use of these building blocks will eliminate much of the tedious component-by-component layout effort and will reduce the risk of errors.

Packaging Requirements

IC packages have die size limitations. Large arrays may not physically fit in a small surface mount or 8-pin DIP package regardless of the number of pins used. Be careful to investigate packaging options for an array you are considering before you start layout. In very space-sensitive applications, you may consider putting a circuit in two small chips as opposed to one large one. Less circuit board area may be required for two small custom ICs in small packages than for one large chip in a large package. This depends in part on the number of external components and how they connect to the IC. Using a large array with poor layout efficiency can easily eliminate packaging options. Check package availability with any vendors being considered as an alternate source. Even if a second vendor offers the same (or an equivalent) array as your primary vendor, he or she may not offer the same packaging options.

6.2 Macro (Chip-Level) Layout

The first step in the layout is to roughly locate major circuit blocks on the die. The position of some blocks will be dictated by the location of specialized components or cells on the die. With these blocks allocated first, the rest of the circuit blocks can be placed in the remaining space.

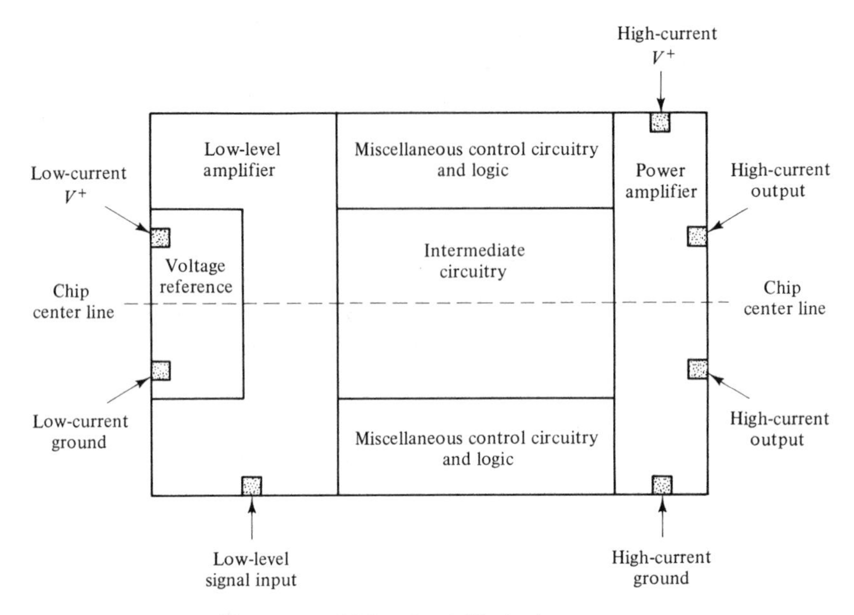

Figure 6-1 Major circuit block placement.

There are a few precautions to follow when locating the major circuit blocks. First, make sure that sensitive circuitry such as input stages, high-gain amplifiers, or voltage references are located on the opposite end of the die as circuitry that dissipates large amounts of power or changes power dissipation during operation. Keep power dissipators located symmetrically about the center line of the chip. Try to keep the most sensitive stages located farthest from the power dissipators and symmetrically placed about the center line. Choose the power and ground pads to minimize the length of metal interconnections with the rest of the chip. If necessary, use multiple power and ground pads. Be sure to consider the flow of signals on the PC board when laying out the chip. Consider not only isolation on the chip but isolation on the PC board as well.

Figure 6-1 shows an example of major block placement on a chip. The block placements were chosen to provide easy signal flow into and out of the chip while maintaining good isolation on both the chip and the PC board.

The following precautions will help to avoid problems when locating the major circuit blocks.

- **Keep outputs and inputs isolated** This prevents noise from the output coupling back into the input. This can be a big problem if phase shifts make this positive feedback. Never place an output next to a noninverting (positive) input. You are asking for trouble if you do this. Remember that there will be pin-to-pin capacitance in the package and trace-to-trace capacitance on the PC board.
- **Keep power supply and ground leads short** Consider multiple power and ground pins. This will ease IC layout, reduce parasitic voltage drops on the chip power and ground metal and allow for off-chip decoupling of input and output stage power. This can improve isolation considerably. Be sure there will be no signals on the ground (*npn*) or power supply (*pnp*) of current sources or voltage references. These signals can modulate bias currents. Custom ICs do not have to conform to any industry pinout. You have complete freedom with the number of pins and pin assignments. Use these options to your best advantage.
- **Metal sheet resistivity is different between layers** Use the same metal layer to interconnect circuitry sensitive to matching on processes with multiple metal layers.
- **Keep input and output leads short** This will prevent or at least minimize undesirable signal coupling and will reduce parasitic voltage drops on high current output lines.
- **Keep the base lines of current sources away from outputs** This helps to prevent current sources from being modulated by high-current outputs.

6.3 Micro (Circuit-Level) Layout

After the array has been selected and the macro (chip-level) layout completed, the micro (circuit-level) layout can begin. The difficulty of this step will depend on the following:

- The array selected
- The layout efficiency required
- Pinout constraints
- Number of metal layers
- Vendor macrocell availability
- Required level of circuit performance

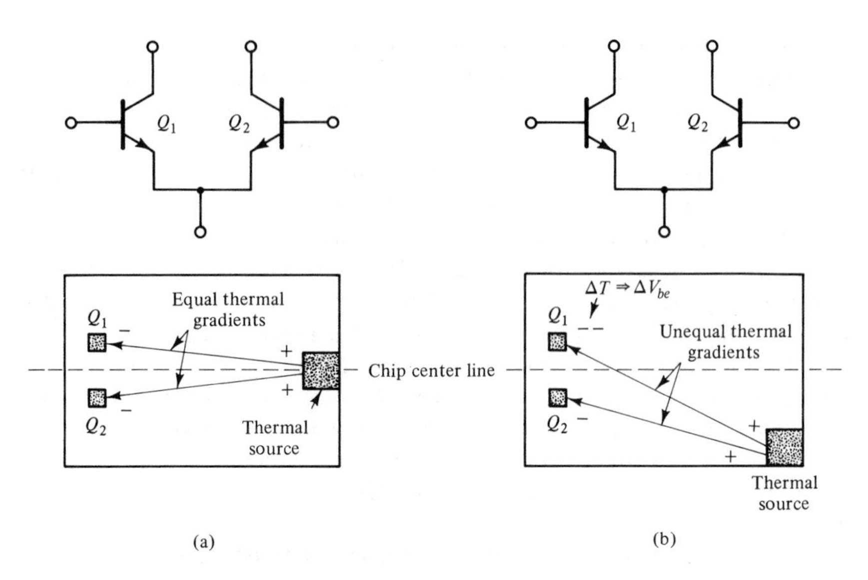

Figure 6-2 Effects of thermal sources. (*a*) "On-axis" thermal source and (*b*) "off-axis" thermal source.

Most of the macro (chip-level) layout techniques apply equally within the microcircuit blocks as they do between the blocks. There are several additional considerations that apply to the detailed circuitry within the blocks. The following guidelines will help avoid many common layout-related circuit functionality problems.

- **Identify thermal sources** Keep sensitive circuitry and input stages away from thermal sources and symmetrically located with respect to the chip center line whenever possible.
- **Cross-couple sensitive inputs** Figure 6-2(*a*) illustrates the effect of an "on-axis" thermal source on a two-transistor differential pair. Since each transistor is located equally from the power source, they are both at the same temperature and no offset is induced. If the thermal source is moved "off-axis," as shown in Figure 6-2(*b*), a temperature differential exists between the two transistors, causing a difference in V_{be} and therefore an offset voltage. Figure 6-3 illustrates how a cross-coupled differential pair can be used to balance the effects of an off-axis thermal source. Each side of the differential pair is affected equally and balance is maintained. This technique is recommended for

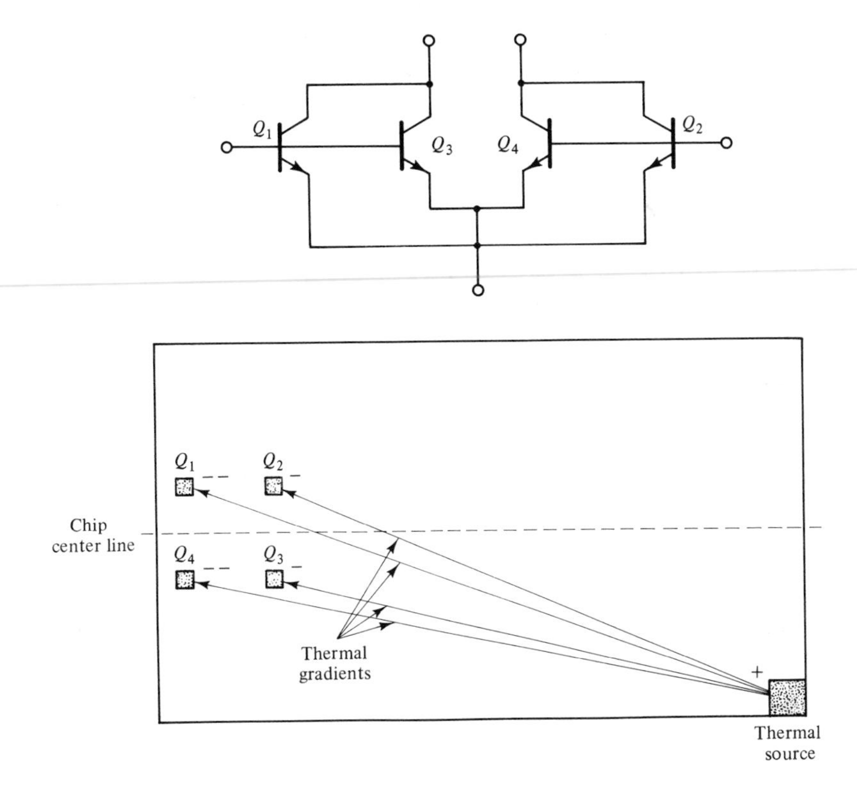

Figure 6-3 A cross-coupled differential stage.

use only in critical applications due to the added circuit complexity and increased component count.

- **Keep matching components close together** It is difficult to match widely separated components due to slight process differences, thermal gradients on the die, and voltage drops in metal lines.
- **Keep matching components oriented in the same direction** Figure 6-4 illustrates the relative orientation of devices for best matching. Devices that are directly adjacent and oriented the same way, as shown in Figure 6-4(*a*), will have the best matching. The components illustrated in Figure 6-4(*b*) will still have "good" matching but not as good as those in Figure 6-4(*a*). The farther apart devices are, regardless of their orientation, the poorer the matching will be. This is due to both thermal- and process-related factors.
- **Collocate current source transistors and reference diodes**

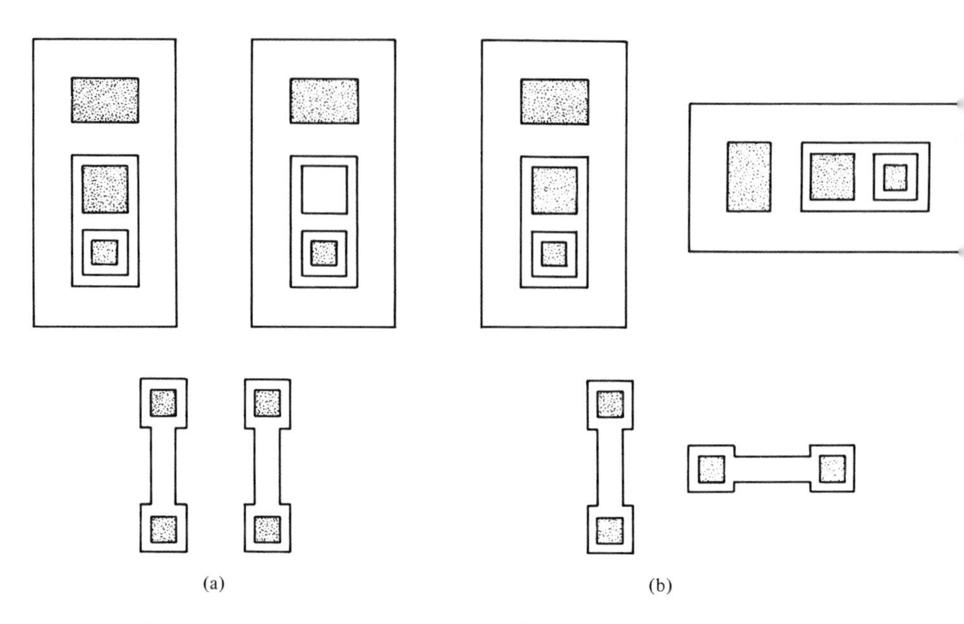

Figure 6-4 Preferred device orientation for (*a*) best and (*b*) poorer matching.

The reference diode for a current-source string establishes the V_{be} on which all of the string currents are based. Thermal mismatches in the V_{be}s of current sources will result in current mismatches. A good layout philosophy is to locate all of the current-source transistors, the reference diode, and any emitter ballast resistors in a compact area and route the collector metal of the current-source transistors to the appropriate locations on the chip. This keeps the bias line for the current-source string short, the current-source components well matched and at the same relative temperature, and minimizes the voltage drops on the metal connecting the current-source emitters.

- **Place cross-unders in high impedance lines** A few hundred ohms will have little effect when connected in series with the collector of a current-source transistor if the current is dc and the voltage on that line does not have to change rapidly. A resistive cross-under in a transistor base or emitter connection or in a high-frequency circuit could be disastrous.
- **Interconnect critical circuitry on the same metal layer** Not only is the sheet resistivity different between metal layers, but the connections between the metal layers can add resistance. These consid-

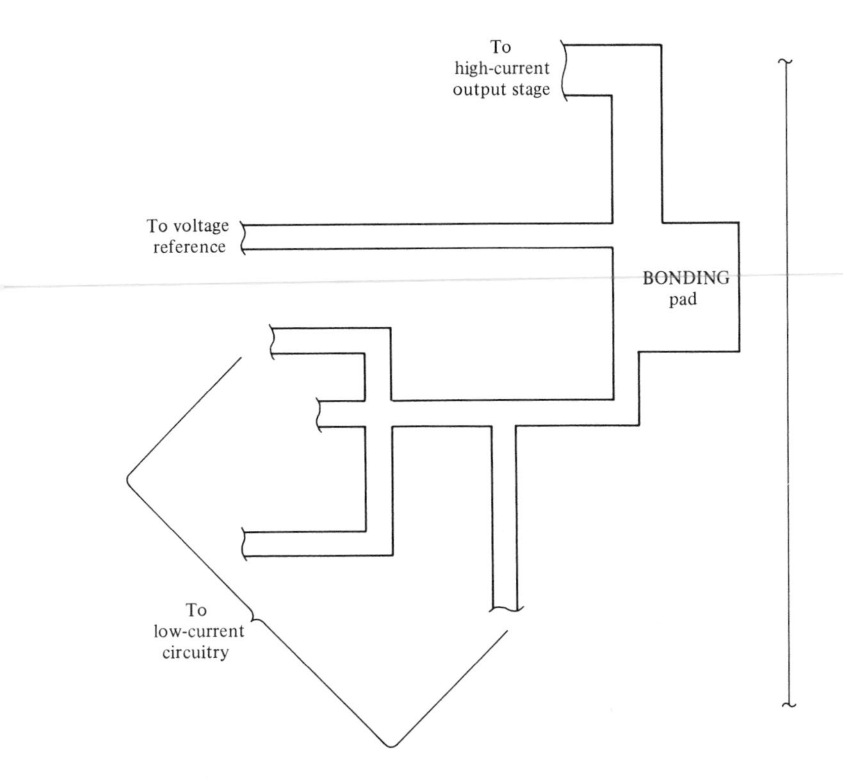

Figure 6-5 Single-point ground.

erations are most important for lengthy interconnections or in cases where the interconnection zig-zags between metal layers.

- **Keep large signal lines, power and ground away from high-gain circuitry** This practice will minimize stray coupling of noise or transients.
- **Use single-point (star) ground for isolation** Figure 6-5 illustrates a desirable grounding technique when separate high-current and low-current bonding pads are not possible. The ideal location for the single-point ground is the ground bonding pad. This single-point connection technique can be used to avoid coupling unwanted signals through resistive drops in the metal lines. The penalty is, of course, added layout complexity. This technique works equally well on power supply and signal lines.

Figures 6-6 and 6-7 illustrate layout sensitivities for typical circuits.

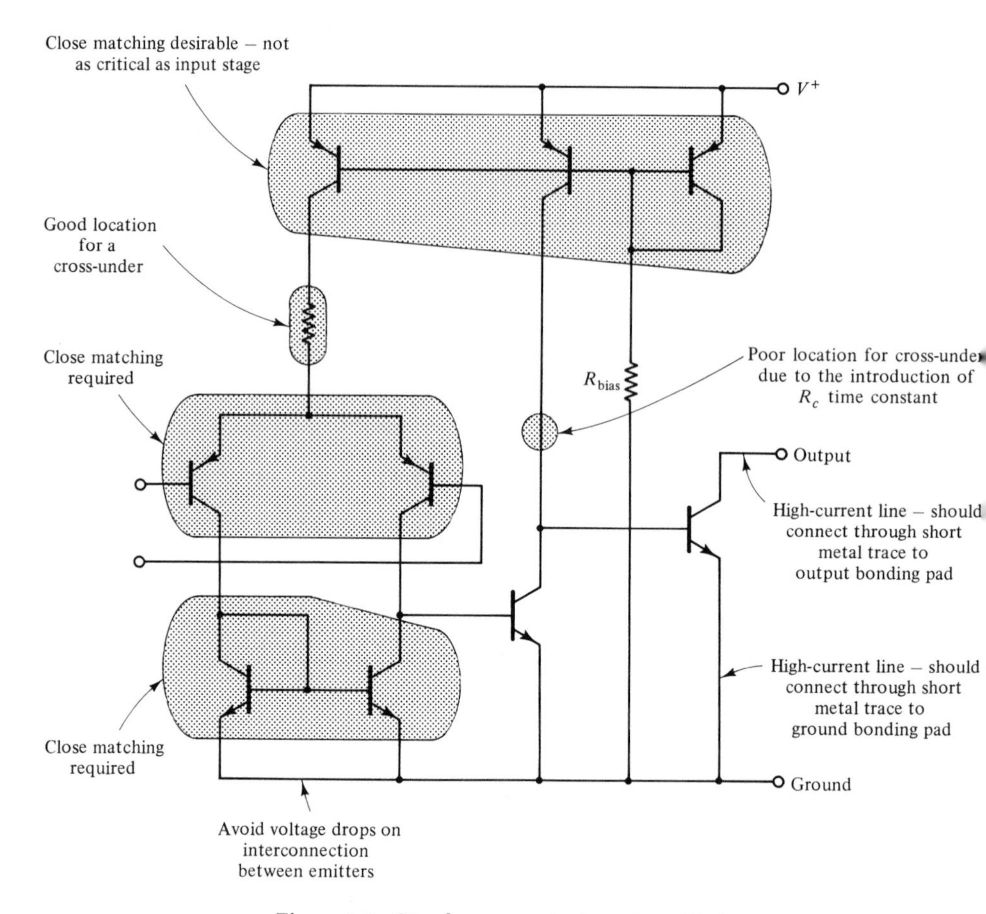

Figure 6-6 Simple comparator layout sensitivity.

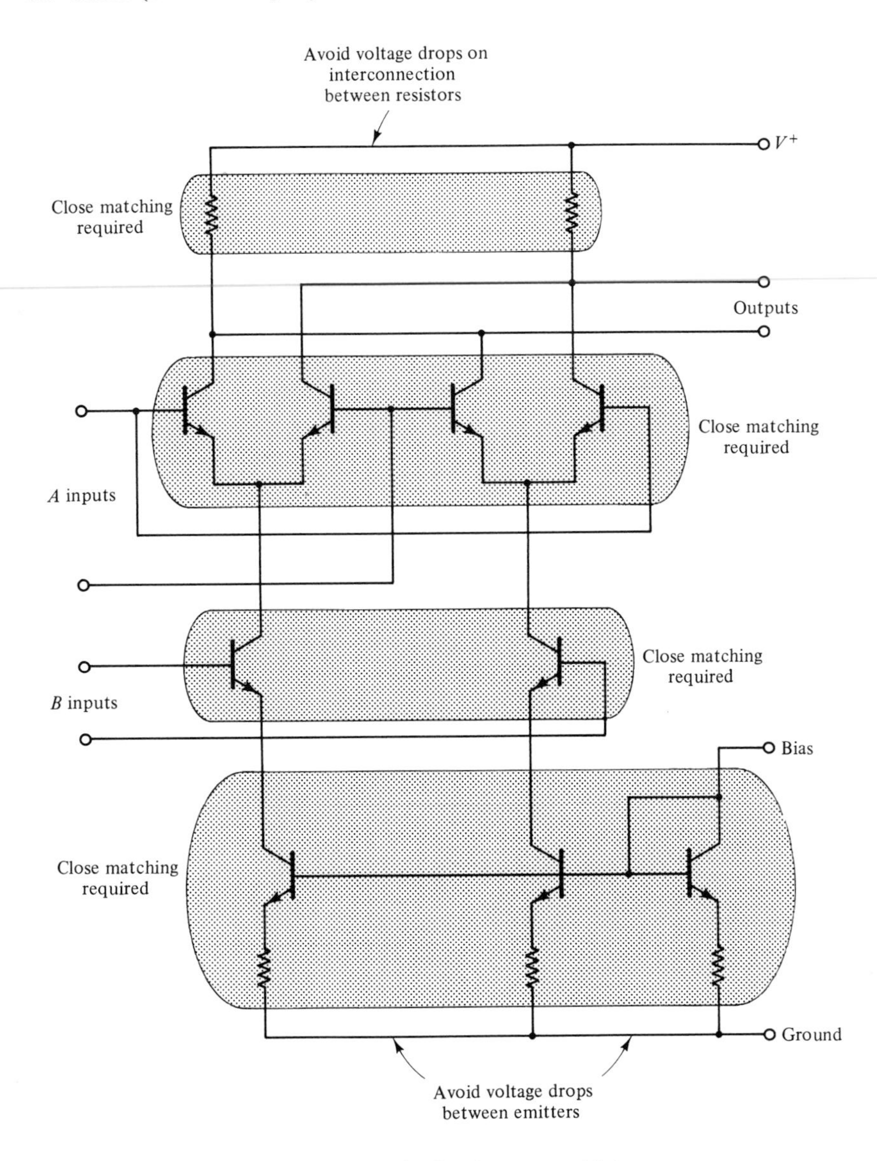

Figure 6-7 Multiplier layout sensitivity.

CHAPTER 7

Packaging

7.0 Introduction

The IC package is as much a component as any transistor or resistor in the circuitry on the chip. Supply voltage, ground, and signals pass between the outside world and the chip through the package. Heat generated on the chip is conducted away and dissipated into the ambient through the package. The chip is isolated by the package from humidity and light. The package also protects the chip and the delicate bonding wires from physical damage.

Selection of the package depends on the specifications of the IC and the environment in which it must function. Some of the most significant criteria used to select the package are

- Number of pins
- Thermal resistance
- Physical size
- PC board technology
- Electrical performance
- Environmental considerations
- Die size
- Reliability
- Cost
- Availability

7.1 Packaging Options

Most ASIC vendors offer a wide variety of plastic and ceramic DIP (dual-in-line) packages. Some vendors offer other packaging technologies that include leaded and leadless chip carriers, SOIC (small outline integrated circuit), and more exotic technologies such as solder bump and TAB (tape automated bonding). Most vendors offer unpackaged die for customers who want to contract their own packaging or use the die in a hybrid assembly. Unless there is a compelling technical reason to use exotic packaging technology, the best choice is to stick with industry standards. Choosing an exotic package available from only one vendor (probably manufactured in very low volume) is the ultimate sole-source risk.

Table 7-1 lists the most commonly available packages. Package specifications vary from vendor to vendor. Get current data from your selected vendor before starting serious design work.

There are several questions to be answered when selecting a package for your analog ASIC. Are there enough pins? You may want to have multiple power and ground pins or to bring out intermediate points in the circuit for testing or alignment purposes. Is the package cavity/mounting tab appropriate for the die size? Is the cavity large enough to accept the die? If the cavity is too large, very long bonding wires may be required. Is the thermal resistance of the package acceptable for the power dissipated on the chip and the ambient conditions under which the chip must function? Will an external heat sink or forced air be required? Are the electrical characteristics of the package acceptable? For example, the pin-to-pin capacitance on DIP packages is the highest on adjacent pins at either end of the package. It is lowest on pins in the center of each side. This could be a serious concern for signals with very fast rise times in high-pin-count (24–40 pin) DIP packages. In these cases, it would be preferable to use the pins in the center of the package for the high-speed signals and the pins on the end of the package for power, ground, and low-speed signals. Chip carriers have low pin-to-pin capacitance on all pins and are worth considering for high-frequency, high-pin-count applications.

Do not forget to consider the cost and availability of sockets and handling equipment. Sockets may be desirable for maintainability, testability, or assembly reasons. Sockets for some IC packages are expensive or unavailable. Also, check the availability of automatic handlers for your test equipment if incoming testing is planned. If you use auto-

Table 7-1
Commonly Available ASIC Packages

No. of pins	Style	Material
8	DIP	Plastic
	DIP	Ceramic
	SOIC	Plastic
14	DIP	Plastic
	DIP	Ceramic
	SOIC	Plastic
16	DIP	Plastic
	DIP	Ceramic
	SOIC	Plastic
18	DIP	Plastic
	DIP	Ceramic
	DIP	Plastic
20	DIP	Plastic
	DIP	Ceramic
	DIP	Plastic
24	DIP	Plastic
	DIP	Ceramic
	DIP	Plastic
28	DIP	Plastic
	PLCC	Ceramic
	LCC	Plastic
40	DIP	Plastic
	DIP	Ceramic
44	PLCC	Plastic
	LCC	Ceramic

matic insertion equipment, be sure the package you are considering is compatible.

7.2 Thermal Design

Power is applied to an integrated circuit which in turn outputs power to the circuitry it drives. The difference between the power applied and the power output is dissipated from the IC as heat. This heat must be removed from the IC while keeping its die temperature within acceptable limits. A given package's ability to do this depends on the power dissipated, the ambient temperature, and the thermal resistance between the IC (junction) and the ambient environment.

Heat flows from the IC to the ambient much like current flows from a current source to ground. While heat flow is analogous to current flow, temperature drops are analogous to voltage drops. They depend on the "thermal resistance" between the heat source and the ambient.

The thermal resistance from the IC to ambient has several components. There is a thermal resistance from the integrated circuit (junction) to case (package), θ_{JC}, and a thermal resistance from the case to ambient, θ_{CA}. Most vendors specify the thermal resistance from junction to case and from junction to ambient. The thermal resistance from junction to case is the most valuable when some sort of external heat sink is used. The thermal resistance from junction to ambient is a measure of the heat-radiating capability of the package and is the most often used. Most vendor specifications for θ_{JA} assume that the IC is soldered into a circuit board. The value of θ_{JA} will be higher if the IC is socketed.

Figure 7-1 illustrates the thermal "circuit" of an IC. P_d is the power dissipated on the chip in watts. This heat flows from the chip to the case and from the case to the ambient environment. Thermal resistance is specified in °C/watt. The thermal resistance allows the number of degrees of temperature rise above ambient for the chip to be calculated per watt of power dissipated.

The thermal resistance between the die and case depends on several factors. The size of the die determines the surface area at the interface between the die and package. The larger this surface area (the larger the die), the lower the thermal resistance between the silicon and the mounting surface on the package. The technique used to mount the die to the package can also impact thermal resistance. Two common methods of die attach are epoxy and gold eutectic. Epoxy is basically a glue, while gold eutectic is a solder connection. Gold eutectic is generally considered a better thermal conductor, but modern thermally conducting epoxies are also very good. Be sure that the thermal resistance number the vendor supplies is valid for the type of die attach that will be used on your production units.

Package lead-frame material (sometimes an option on plastic packages) can have a significant impact on thermal resistance. Typical lead-frame material is either an alloy called Kovar® or copper. Some vendors offer optional copper slugs molded into plastic packages to further im-

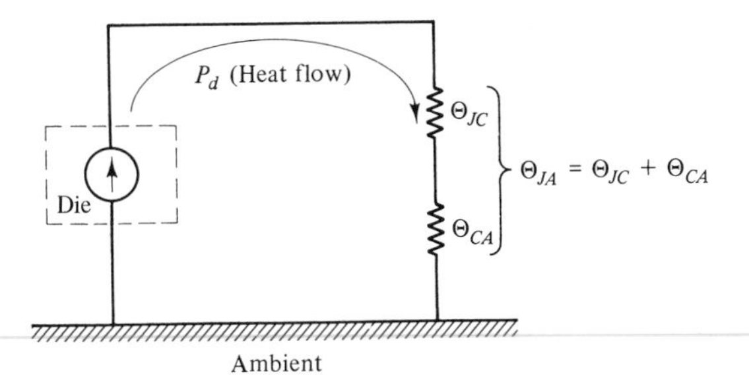

Figure 7-1 The thermal "circuit" of an IC.

prove heat dissipation. Consider the cautions about sole-source packaging technology before making commitments to technically neat but vendor-unique package options. Always consider the impacts should an alternate source become necessary. Higher lead-count packages have lower thermal resistance due to a larger die attach pad, larger package surface area, and more metal pins that serve as mini heat radiators. Check with the IC vendor to get accurate data since thermal resistance depends on several vendor-specific details.

Example 1: Calculate the worst-case junction temperature for a die packaged in a standard 16-pin plastic DIP package. The maximum ambient temperature is 70°C in still air. The die dissipates 250 mW and θ_{JA} is given by the vendor to be 90°C/W. Figure 7-2 illustrates the thermal circuit for Example 1.

The die temperature is

$$T = 70°\text{C} + (90°\text{C/W})(0.25\text{ W}) = 92.5°\text{C}$$

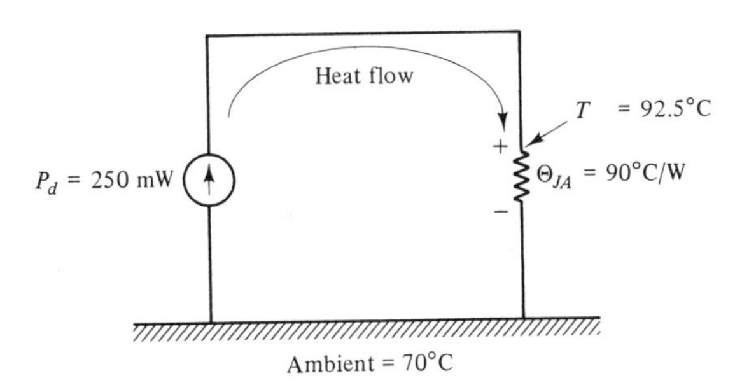

Figure 7-2 Thermal circuit for example 1.

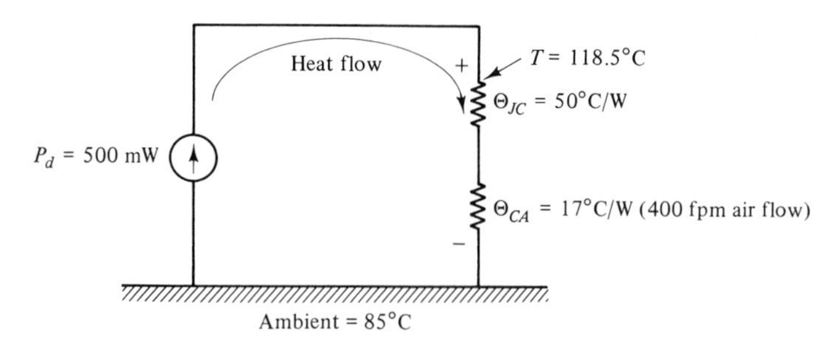

Figure 7-3 Thermal circuit for example 2.

Example 2: Calculate the worst-case junction temperature for a die packaged in a standard 24-pin plastic DIP package with a Thermalloy 6085B heat sink attached to the package with Thermalbond epoxy. The IC dissipates 0.5 W, the maximum ambient temperature is 85°C, and air is flowing at 400 fpm. The ASIC vendor specifies θ_{JC} as 50°C/W and Thermalloy specifies the thermal resistance of the heat sink as 17°C/W. Figure 7-3 illustrates the thermal circuit for Example 2.

The die temperature is

$$T = 85^\circ\text{C} + [(50 + 17)^\circ\text{C/W}](0.5\ \text{W}) = 118.5^\circ\text{C}$$

Thermal resistance depends on die size, die attach method, and package characteristics. The curves in Figures 7-4 through 7-6 illustrate θ_{JA} versus package pin count for plastic DIP, ceramic, and SOIC packages.[1] These curves are meant to reflect typical numbers and should be used for comparison only. Obtain current data from the ASIC vendor you are considering for the package and array to be used before proceeding with detailed design.

Die temperature has a significant impact on reliability. Most IC vendors list the absolute maximum die temperature as 150°C. In a military application with $T_{A_{max}} = 125^\circ$C, there can be only a 25°C temperature rise on the chip over ambient (not considering derating requirements). Although chips are specified to operate at temperatures up to 150°C, failure rate increases rapidly at elevated die temperatures. Figure 7-7 plots failure rate versus die temperature, assuming an activation energy $e_a = 0.99$ eV.[2]

$$\text{Failure rate} = \exp\left[\frac{e_a}{k}\left(\frac{1}{T_2} - \frac{1}{T_1}\right)\right]$$

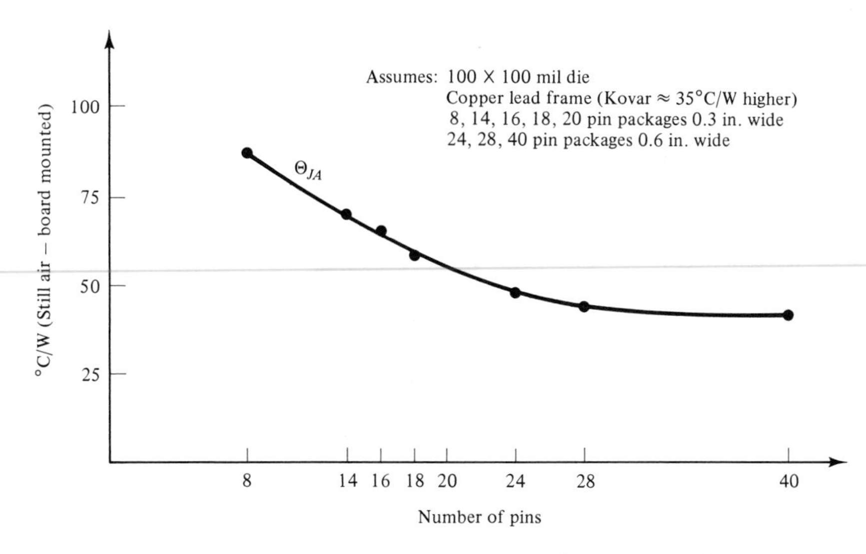

Figure 7-4 Approximate plastic DIP thermal resistance (Ref. 1).

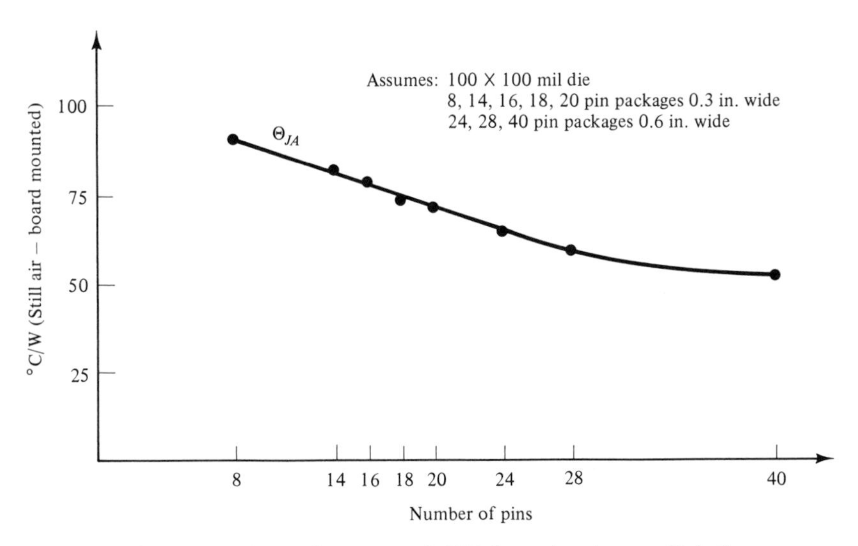

Figure 7-5 Approximate ceramic DIP thermal resistance (Ref. 1).

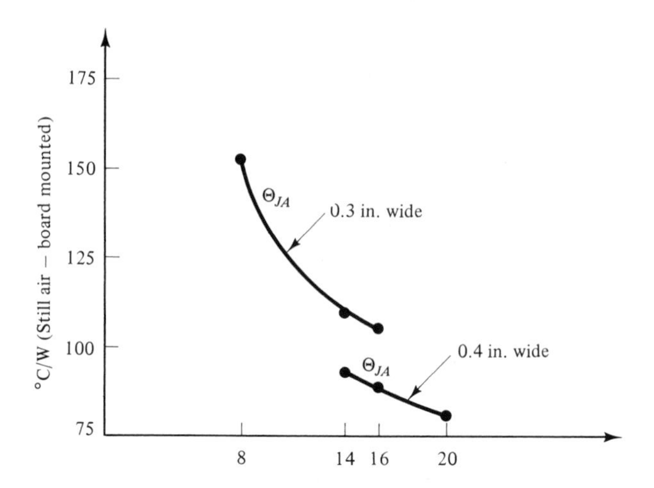

Figure 7-6 Approximate plastic SOIC thermal resistance (Ref. 1).

where

$$\begin{aligned} e_a &= \text{activation energy in eV} \\ k &= \text{Boltzmann's constant} \\ T &= \text{die temperature in Kelvins} \end{aligned}$$

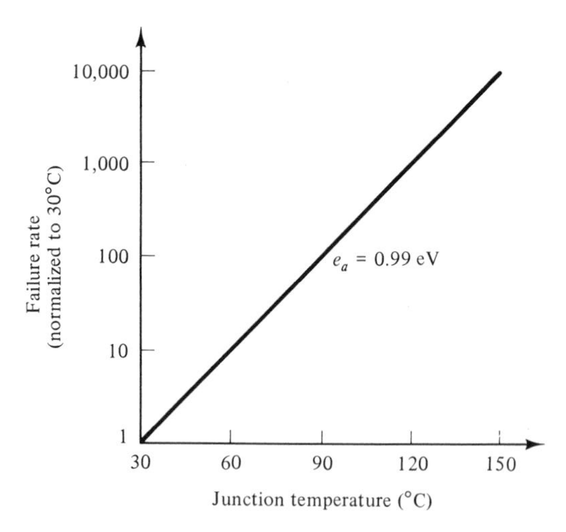

Figure 7-7 IC failure rate versus die temperature.

For each 30°C rise in die temperature there is an order of magnitude increase in failure rate. Keep die temperatures as low as possible to minimize failure rates. Use external components to dissipate power whenever possible. Power transistors, for example, are available in packages better suited for heat sinking than high-pin-count ICs.

7.3 Package Selection

Final package selection is an important part of the ASIC design and should be given as careful consideration as any other part of the design. Preliminary package selection will be made during the early stages of IC definition. However, a thorough review of the package must be included in the design finalization. This step is important. Costly and time-consuming redesign of the IC or end product can result if an error in package selection is made.

The following list is a "package selection checklist" that covers most (but not every possible) consideration important in choosing the optimum package.

- **Pin count** Does the package have an adequate number of pins? Can the IC be laid out to make an effective and efficient use of the pins? Will the layout be unnecessarily complicated in an attempt to use a low-pin-count package? Has consideration been given to bringing intermediate circuit points to pins for testing, alignment, and troubleshooting purposes? Careful planning and a slightly higher package cost could save a significant amount of money in testing, manufacturing, and rework and repair. Consider partitioning the custom IC into two small arrays in two low-pin-count packages. Will the ICs be easier to test? Will the two chips be less expensive? Will the PC board in the end product be easier to lay out? Will the resulting circuit flow on the PC board be better from a performance perspective? Will the power dissipation be partitioned between the two chips such that a potential thermal problem will be avoided?
- **Package size** Will the package physically fit in its intended location? For example, some ASIC vendors offer 0.3-in.-wide 24-pin plastic DIPs. Most offer only 0.6-in.-wide 24-pin plastic DIPs. Be sure you specify the package size, not just the pin count and package type. A package outline drawing with dimensions should be included as part of your formal specification.

• **Cavity size** Be very sure you know well in advance what the die size limitations are for any package under consideration. Nothing is more disappointing than to count on using a small package like a 16-pin SOIC only to discover when the circuit design and IC layout are complete that the array won't fit into the desired package and a much larger alternative must be selected. Many engineers unfamiliar with IC technology casually assume that chips are "small" and will fit in any package. This is not the case.

• **Mounting technology** Have all of the implications of the selected mounting technology been fully investigated? Surface-mount packages can save a tremendous amount of circuit board area but require different assembly techniques than conventional through-hole technology. Are there any hidden costs for handling, test, or assembly equipment? This is especially true for high-volume applications. Leadless ceramic chip carriers can cause unexpected problems if the PC board or substrate they are soldered to has a different coefficient of expansion over temperature than the LCC. Cracked packages, broken solder joints, or separated PC board traces can result since there are no leads to absorb the thermally induced stress. Leaded chip carriers solve this problem but can cause other assembly-related problems if the leads are not coplanar. This problem is exhibited if the IC is placed on a flat surface and not all of the leads touch the surface. The leads can become noncoplanar during device assembly, testing, or subsequent handling. Solder contact with one or more pins may not be made if the lack of coplanarity is significant enough. If you plan on using sockets, be sure they are available within the lead time and cost limitations you expect. Also, be sure that the sockets selected meet your size and height requirements. Some sockets are larger than might initially be anticipated.

• **Thermal characteristics** Calculate the maximum power the chip will have to dissipate. Use the maximum power supply voltage and maximum signals applied to the chip or driven by the chip. Determine the real worst-case ambient temperature for the chip. If the chip is in an enclosure, the ambient is the temperature inside the enclosure, not the temperature to which the enclosure is exposed. If the IC is positioned next to a significant power dissipator such as a voltage regulator, dropping resistor, or a large current output transistor, the local ambient for the IC could be significantly higher than the average ambient for a piece of equipment. If you plan to use heat sinks, be sure that they are adequate, available, and meet your cost size and height requirements. The IC and heat sink must be mounted on the PC board such that the heat sink can efficiently operate to the desired specifications.

In other words, do not obstruct the flow of air past the heat sink or place it next to a large heat source and expect its thermal resistance to be unaffected. Remember that junction-to-ambient thermal resistance numbers assume the device is soldered directly into the PC board. If the IC is socketed, the thermal resistance will be higher.

- **Miscellaneous considerations** Some IC packages have metal surfaces. Most of the time these are at substrate potential. Problems will result if these metal surfaces touch PC board traces, feedthroughs, other components, chassis, or shielding material. Heat sinks used with these types of package will be at the same potential as the package. Very high pin count packages such as pin grid arrays and large DIPs can have an amazingly high insertion and removal force even in so-called low-insertion-force sockets. Be sure it is advantageous to consider sockets in these cases and allow room for any necessary mechanical extraction device that may be required. Consider alternate sources and future availability of the package.

References

1. *Linear Databook Volume 1*. National Semiconductor, Santa Clara, Calif., 1988; pp. 7–22.
2. *Linear Databook Volume 1*. (1988), pp. 7–18,19.

CHAPTER 8

Testing

8.0 Introduction

Quality is the measure of a device's ability to initially meet a set of specifications. Reliability is the measure of a device's ability to meet these specifications over time. Testing is used both to force quality and reliability and to measure and monitor it. Typically, a product is subjected to operational testing using test limits tighter than the formal specifications. This is an attempt to eliminate any marginal units which could result in initial or long-term failures. Additional testing of devices that satisfactorily pass the first round of testing would be used to monitor the results. These secondary test results could be used to modify the initial screening process to optimize the yield while still guaranteeing a specified level of quality and reliability.

Ideally, the required level of quality and reliability are designed into a product through good circuit design practices and an adequate margin in the design. In this case the testing process will be attempting to uncover overt manufacturing defects, which affect quality, and latent manufacturing defects, which affect reliability.

A typical product (in this case an analog ASIC) will have a failure rate that follows the curve illustrated in Figure 8-1. The initially high failure rate is due to latent manufacturing defects. Most manufacturers issue warrantees for products such as automobiles, stereos, and computers to cover product failures that surface during this period. This

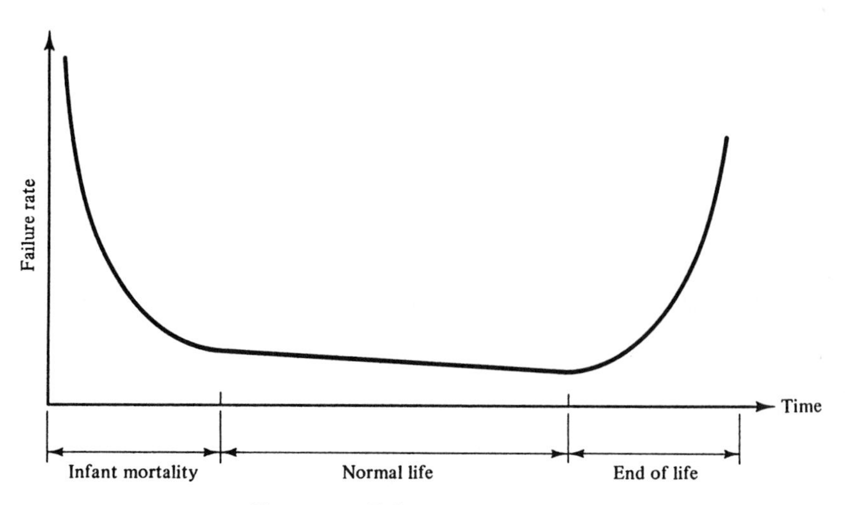

Figure 8-1 Failure rate versus time.

failure rate can be very significantly reduced by rigorous testing. This testing, however, is expensive. A balance must be struck between the cost of warranty repair and the cost of testing. The amount of testing depends on the application of the device. If an analog ASIC is to be used in a satellite, for example, a great deal of testing would be warranted since repair costs would be astronomical (pun intended). This would not be the case for an inexpensive telephone.

The flat portion of the failure rate curve is the period of normal product life. During this time the failure rate (for an adequately designed part) is low. There are event-related failures and contributions from manufacturing defects that outlive the infant mortality period and some early wear-out failures. There is probably little difference between the last two failure mechanisms.

The final period of a product's life is the "end of life" period. This is when the product is "worn out" or "used up." The wear-out period for integrated circuits usually occurs long after the product using the IC is outdated and is not typically a concern except in more exotic applications.

Testing done on integrated circuits can be broken down into two categories: electrical and environmental. Electrical testing measures the dc, ac, and time-domain performance of an IC. Environmental testing measures the electrical performance and mechanical integrity of a device over a wide range of environmental variables such as temperature, shock, vibration, humidity, and mechanical stress. The amount of testing and the severity of the tests are determined by the quality and reliability goals for the IC and the end product in which the IC must ultimately function.

8.1 Electrical Testing

The goal is to test as little as possible and still guarantee the required level of performance (quality and reliability). Think in terms of the overall system and the ASIC characteristics, specifications, and test methodology necessary to meet the system requirements. Remember, the goal of the ASIC development was to design a more profitable system, not to design a tightly specified IC.

There are two costs associated with the testing activity. The first cost is for the initial development of the test program and fixtures to be used on the vendor's automatic test equipment. Second, there is the recurring cost associated with production test time. Simple effective tests minimize both of these costs.

The ASIC vendor performs several layers of electrical tests on the IC as it is manufactured. The sheet resistivity of diffused layers is measured as the wafer is processed. Breakdown voltages and transistor beta are also measured. These measurements ensure that wafer processing has been done correctly and that process specifications are met. Next, newly assembled devices are tested for package pins shorted during the assembly operation. Finally, functional testing is performed. Functional electrical testing can be broken down into three different categories. These are, in order of increasing difficulty and expense,

- dc testing
- ac testing
- Time-domain testing

The electrical test plan, including test methods and acceptable test results, should evolve with the design. Tests are a natural part of the circuit design process. Tests are performed during computer simulations with SPICE and while evaluating a breadboard. All that is necessary is to document these tests, conditions, and desired results. If this is done during the design phase, you will be a long way down the road toward having a highly valid test procedure defined when the design is complete.

While the final test plan is being written, keep in mind that semiconductor manufacturing is a high-volume business. Everything is as automated as possible. Therefore, any exceptions to the normal production flow will add considerably to the cost of an IC. Most production IC testing is done with dc or low-frequency (<1 MHz) ac measurements. Any other testing requires special modifications to normal procedures, handling equipment, or very expensive hand-testing by a technician. If an ac or time-domain test is specified, most vendors will attempt to meet this requirement through correlating the results of a dc test to that of the desired ac or time-domain test. For example, it may be possible to correlate dc open-loop gain or power supply current with bandwidth and rise time. This way, a simple power supply current measurement (which is done anyway) or dc gain test can be substituted for a much more complex test. This testing by proxy can save a great deal of time and therefore money. The vendor will want things to work out this way. However, you must be certain the correlation is valid before approving this as part of the test plan. Otherwise, you may be on the receiving end of a great deal of product that may not meet your needs.

Once production deliveries of the ASIC begin, it is possible to receive units that fail an incoming test or fail to function correctly in the product. This might happen initially or several years in the future. This can be caused by one or more of several things. The production tests may not cover all parametric variations that can render a part unacceptable. The vendor's process may change. The base sheet resistivity may be increased or decreased slightly to improve overall production yield. If your device is sensitive to this type of change and production testing is unable to trap the unacceptable devices that result from these changes, the reject rate can increase sharply. The vendor may purchase a new tester and attempt to translate your test program to the new machine. If something is lost in the translation, the result could be bad parts.

The severity of these correlation problems can be minimized with "standard parts." Serialize and thoroughly characterize 6 to 10 of the initial functional prototypes. Keep these parts and their data safe. When a correlation problem comes up, these "standard parts" with their known data can be used to trace the source of the problem. Periodically characterize several current production units. The data will allow you to monitor the quality of the incoming product. If the parts are serialized and kept with their data, you will have a continuous supply of "standard parts." This process also allows you to detect any shift in performance of the devices and may alert you to a potential problem. You will then have time to work with the vendor to mitigate any risk to your production.

8.2 Environmental Testing/ Quality Assurance

A great number of tests and inspections are routinely performed by an IC vendor during his normal production flow. Some of these tests are destructive and must be done on a sample basis. Other tests are not destructive but are done on a sample basis for cost reasons. Still other tests and inspections are done on every device. Most IC vendors have several standard "process flows" through which parts can be directed. They typically include separate flows for commercial-grade devices, industrial-grade devices, high-reliability devices, and several military flows. Normally the commercial flow is quoted unless otherwise specified. The more rigorous the process flow, the more expensive it will be.

The purpose of these tests and inspections is to catch any failures or potential failures as early in the manufacturing process as possible. There are few economically practical methods of reworking semiconductor material, and defective material is therefore usually scrapped. At each step throughout the manufacturing process from wafer fabrication through assembly and final test, the cost to scrap a device grows. It is in everyone's best interest to scrap flawed material as early as possible. Higher production costs impact the vendor's profit and will eventually be passed on to the customer.

Table 8-1 illustrates typical process flows followed for various types of devices. Most vendors are willing to perform special testing, such as temperature testing on commercial plastic parts, for an added charge. The customer will have to weigh the value of these tests versus their costs.

Device failures can be divided into four broad categories: design failures, process failures, assembly failures, and testing failures. The inspections and tests in the process flows have been designed to detect the mechanisms that cause failures in each of these categories. Some of these tests and inspections are done routinely, others are done on a lot sample basis, and others are optional or are done only in high-reliability or military manufacturing flows.

Table 8-1
Typical Process Flows

Test/inspection	Commercial	Industrial	High reliability
Wafer sort	100%	100%	100%
Die visual	100%	100%	100%
Preseal visual	100%	100%	100%
Die attach	N/A	N/A	Sample
Bond strength	Sample	Sample	Sample
Stabilization bake	N/A	100%	100%
Temperature cycle	100%	100%	100%
Const. acceleration	N/A	N/A	100%
Leak (fine and gross)	N/A	Sample	Sample
Electrical test (25°C)	100%	100%	100%
High-temp test	N/A	100%@85°C	100%@125°C
Low-temp test	N/A	N/A	100%@ − 55°C
Burn-in	N/A	N/A	100%
Post-burn-in elect	N/A	N/A	100%
External visual	N/A	N/A	100%

Design Failures

Design failures are due to the marginality of a circuit implementation or a process for the particular application in which it is being used. The only ways to eliminate these failures are to test under the conditions of most marginality (high temperature, low supply voltage, etc.), redesign the device, or implement the current circuit on a more robust process. Devices in this category may require minimum transistor f_ts that are process nominals, or process breakdown voltages may be marginal for the application such that degradation occurs over time.

Fabrication Failures

These failures are due to faults in the wafer fabrication process. There may be pinholes in the oxide layer, incorrectly defined diffusions due to a dust speck on a mask, or chips in the die or oxide. The metallization may have scratches, voids, or shorts. There may be ionic contamination or passivation on the bonding pads.

Assembly Failures

Assembly-related failures are due to defects in the process of placing an otherwise good die in a package. These problems include defects in the bonding wires such as weak bonds, poor die attach, and lack of mechanical integrity of the package.

Testing Failures

Testing failures are caused by tests not being performed, being performed incorrectly, or guard bands not wide enough to allow for marginal results.

Table 8-2 lists the four categories of failures and the procedures used to detect the responsible failure mechanisms.

The following are brief descriptions of the test procedures and inspections typically performed during normal integrated circuit process flows. These descriptions are generic. The ASIC vendor(s) you have selected should be consulted for a detailed description of the procedures they follow. It is important to understand what level of screening your devices will receive. Make no assumptions.

- **Wafer sort** Wafer sort is a simple electrical test performed on each die on the wafer. Reject die are marked with an ink drop so they can be discarded after the die are separated. This is the first electrical test performed on the completed die and is designed to reject gross failures.
- **Die visual** After the rejected die are removed, the remaining electrically good (at the wafer sort level) die are visually inspected to check for scratches in the metallization, chips or cracks on the edge of the die, and passivation on the bonding pads. Any die that fail this inspection are discarded.
- **Preseal visual** Once the die is mounted in the package and the bonding wires are attached, the connected IC is visually inspected before the package is sealed or encapsulated. This inspection checks for faulty wire bonds, faulty die attach, chip damage caused during the assembly process, and foreign material in the package cavity or on the lead frame.
- **Die attach** A sample of an assembly lot is taken to test the strength of the die attach. A tool is used to apply a shear force on the die in an attempt to dislodge it. This is a destructive test to mea-

Table 8-2
Test and Inspection Procedures

Test inspection	Design	Fab	Assembly	Testing
Wafer sort	•			•
Die visual		•		
Preseal visual		•	•	
Die attach			•	
Bond strength			•	
Stabilization bake		•		
Temperature cycle		•	•	
Const. acceleration			•	
Leak (fine and gross)			•	
Electrical test (25°C)	•	•	•	•
High-temp test	•	•		•
Low-temp test	•	•		•
Burn-in		•		
Post–burn-in elect	•	•		•
External visual			•	

sure the integrity of the die attach. Failures are usually due to voids under the die or a poor-quality bond between the die and the mounting surface of the package. Defective die attach causes an increase in thermal resistance from the die to the package (and to the ambient), resulting in elevated die temperatures. In the worst case, the die can become loose from the package mounting surface and literally be suspended by the bonding wires.

- **Lead bond strength** A sample of an assembly lot is taken to test the strength of the wire bonds. A tool is used to pull on the bonding wire in an attempt to dislodge it. This is a destructive test to measure the integrity of the wire bond. Failures are usually due to glass on the bonding pads or a poor-quality bond between the wire and the bonding pad.
- **Stabilization bake** The integrated circuits are baked in an oven at 150°C without power applied. This allows the redistribution of mobile ionic contamination on the chip and is done prior to electrical testing.
- **Temperature cycle** The ICs are alternately heated and cooled without electrical power applied. This thermal stress will cause a marginal package to break and cracks in faulty passivation on the chip or a faulty die attach to separate.

• **Constant acceleration** The ICs are placed in a centrifuge and subjected to 30,000 times the force of gravity. This test checks the integrity of wire bonding, packages, and die attach and is performed on devices with nonencapsulated packages.

• **Leak test (fine and gross)** This test is performed on devices in ceramic cavity packages without power applied. The purpose of these tests is to check package integrity and eliminate devices that might fail in high-humidity environments. For the fine leak test, devices are placed in helium under high pressure. Helium will seep into the die cavity through any cracks or pinholes in the package. The devices are next put into a vacuum chamber and the rate of helium leaking out of the package is measured by a mass spectrometer. The gross leak test is similar to the fine leak test except that, after a pressure treatment, the devices are submerged in a heated liquid and a visual inspection is made for bubbles coming from the package.

• **Pre–burn-in electrical test** This is a complete functional test to final specifications at 25°C. If no burn-in is to be performed, this would be the final electrical test. The ICs would be serialized prior to this test if parametric data is to be recorded for this and subsequent tests.

• **High- and low-temperature electrical tests** These are complete functional tests to final specifications at the specified high and low temperatures. If no burn-in is to be performed, these would be the final electrical tests. The ICs would be serialized prior to 25°C testing if additional parametric data is to be recorded for these tests. In most cases these tests are go/no-go tests and no data is taken unless the data is used for characterization.

• **Burn-in** The devices are electrically stressed and heated to 125°C for a specified number of hours. This process accelerates the failure rate, causing marginal devices to fail during the burn-in period.

• **Post–burn-in electrical test** This is a complete functional test to final specifications at 25°C. This test eliminates any devices that fail or have parametric values that shift outside acceptable limits as a result of burn-in.

• **External visual** This final visual inspection verifies that the package leads and marking are free of defects.

Testing is a tool used to enforce and monitor both quality and reliability. When testing is used in the proper measure, it is a very cost-effective step in the manufacture of an analog ASIC.

CHAPTER 9

Custom IC Economics

9.0 Introduction

There are many reasons why a custom IC would be contemplated. It may be necessary to reduce the size of a product or lower production and inventory costs. A custom IC is more difficult and expensive for a competitor to copy. Custom ICs are more reliable than discrete designs and make products easier to troubleshoot. Production testing is simplified since custom ICs represent a large portion of a product's circuitry and come pretested to the customer's specifications. A product with an "integrated circuit" may have a marketing advantage over competing products. All things considered, custom ICs are attractive for one reason: They offer financial advantages. The success of a custom IC development is measured by the magnitude of the financial reward it delivers.

The purpose of this chapter is to examine the economics of custom ICs. Guidelines for defining custom ICs and determining the cost effectiveness of the integrated approach are presented. A thorough "up-front" analysis of the economic considerations of a custom design is essential to a successful development effort. This financial window can give a valuable insight when making decisions and trade-offs throughout the design and vendor-selection process.

9.1 Defining the Custom IC

Defining the function and performance of a custom IC requires careful risk analysis from both a technical and economic point of view. On one extreme, a custom IC may be technically simple and have a significant positive economic impact on a product. The other extreme would be a custom IC with significant technical risk and marginal potential financial benefit. In these cases, once the technical and economic facts are understood, the decision of whether or not to proceed with the custom IC development requires little thought. Most cases, however, lie somewhere between the extremes. It is not readily apparent whether or not the benefits outweigh the risks or vice versa. Figure 9-1 illustrates that, as economic risks increase, the potential rewards must offset the risk for the custom IC to be within the realm of consideration. The slope of the dividing line between "worth consideration" and "not worth consideration" depends on the decision maker and the amount of risk he is willing to take.

Technical Risks

• **Pushing the performance capabilities of an IC process** Bumping against process performance limitations can make production yields marginal, resulting in higher unit prices and delivery problems. Detailed and exotic testing may be required to guarantee compliance with specifications, leading to further cost and delivery impacts. Difficulty correlating test results may lead to disputes with the vendor over whether or not parts meet specifications. Parts the customer rejects at incoming inspection may pass the vendor's production testing. Specifications involving frequency response, delay times, noise (voltage or current), offset (voltage or current), and absolute values (voltage, current, resistance, etc.) can lead to trouble if they are marginally achievable with the process or device geometries available.

These risks can be reduced by careful characterization of the process and device geometries to ensure accurate device models, detailed simulations considering process variations, and careful up-front definition with the vendor of test conditions and methods.

• **Using a technology to perform nonoptimized functions** Attempting to perform high-density logic functions on a linear bipolar array or precision analog functions on a CMOS gate array are high-risk

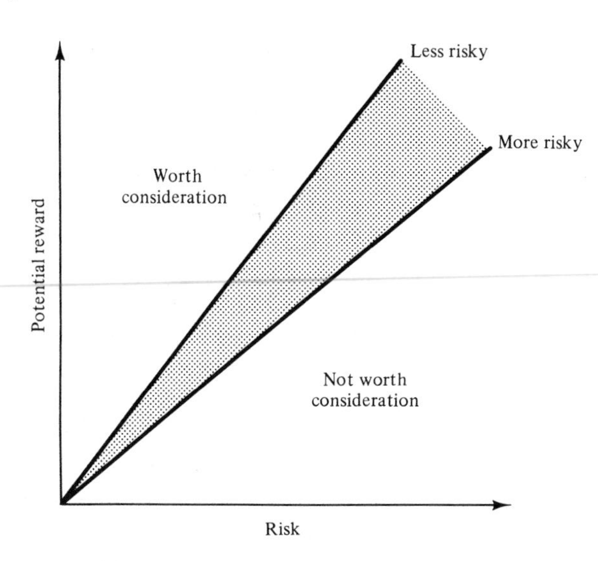

Figure 9-1 Risk versus reward trade-off.

propositions. Linear bipolar arrays do not typically have the transistor density necessary to perform high-density logic. The numbers and values of resistors on these arrays are not optimized for gate design. The currents required for these logic circuits can become very large. CMOS arrays are typically designed to implement high-density logic. They have very few resistors or capacitors. Most of the transistors have very small geometries and short channel lengths. These devices are not well suited to analog designs.

Partitioning the system to optimize performance through proper choices of technology will minimize this risk. In cases where it is necessary to cross technologies and functions, be sure the capabilities and limitations are well understood before proceeding. An alternate approach to the circuit function may be worth considering.

• **Choosing a technically unique sole-source product** Integrating an ASIC on an array with characteristics such as very high breakdown voltage, thin film resistors, or unique component geometries increases risk. These types of arrays are usually offered by a single vendor. The risk is reduced if your circuit can be readily adapted to a different array offered by the same or another vendor. The largest risk is the lack of availability of the array. This could occur with either production problems or the manufacturer discontinuing the array due to low demand. These concerns also apply to new unproven products and exotic packaging technology. These risks can be minimized by partitioning the system and implementing the circuitry to eliminate the dependency on exotic technologies and unproven products.

- **Attempting to achieve an excessively high level of integration** In an attempt to use "every device" on an analog array, layout compromises will almost necessarily occur. Just because a circuit is "connected" correctly does not mean it will function correctly. Thermal feedback, noise coupling, and voltage drops on long bias, ground, and power supply lines can cause a circuit to malfunction. A small change to the circuit may require a complete relayout of the chip due to the lack of conveniently located spare devices. In the extreme, it may be necessary to redesign portions of the circuitry to reduce the number of components, thus creating spares or eliminating the need for additional devices. Ultimately, the next larger array may be required. A good rule of thumb is to count on using no more than about 80% of the components on an array. The useful number of components will vary, depending on the array style and the circuit.
- **Designing without well-defined specifications** Many times, assumptions made at the start of a design are not revised when formal specifications are established. These hidden flaws can torpedo an otherwise brilliant design. This risk can be minimized by revalidating all assumptions and design approaches every time a specification or requirement is changed.
- **Using inexperienced design resources** No matter whether the ASIC vendor's engineers, the customer's engineers, or an independent design service designs the ASIC, there will be some level of inexperience. The ASIC vendor or independent design service may have intimate knowledge about designing ASICs but will have little experience with the customer's product. The customer's engineers, on the other hand, will have a thorough knowledge of the product but will never have designed an analog ASIC before. The decision of which resource to use will depend on schedule pressures, economics, and the availability, experience, and eagerness to learn of the customer's engineers. If the customer wishes to develop additional analog ASICs in the future, it may be worth developing an in-house design capability. If an outside resource is chosen to perform the design work, it is critical that close working relationships exist between the customer's engineers, the IC design engineers and the vendor's engineers. Undoubtedly, some ASIC/system trade-offs will have to be made during the course of the design effort.

Economic Risks

- **Development cost and risks are not offset by achievable financial benefits** This risk can be mitigated by a thorough understanding of development cost (all costs), achievable benefits, and risks up front.
- **Product market window too narrow** Design errors and resulting rework cycles can delay the introduction of the new product, thereby reducing the payback of the custom IC. If the market window is too narrow, a delay in introducing the custom IC version of a product may remove any financial benefit of the custom IC.
- **Design iterations can add unexpected cost** Design iterations made necessary by requirement changes or design errors will cause the expenditure of additional engineering time and can result in additional layout and prototype integration charges by the vendor.
- **Price increases in the unit price of the IC** Unit price increases may result if the final IC requires more-expensive packaging, more-exotic testing, or the final specifications result in a lower manufacturing yield than was originally anticipated by the vendor.
- **Single-source supplier.** There is nothing inherently wrong with having a single-source supplier. There are, however, some risks. Single-source suppliers are less likely to negotiate pricing (where else can you go?). Parts delivery may be delayed if the vendor is working near capacity or receives an unanticipated large order; if there is no other source for the product, the delays must be tolerated. Deliveries could be completely interrupted if the vendor suffered a disaster such as a fire, hurricane, or earthquake. Risks to deliveries can be mitigated by purchasing parts ahead. This, however, increases inventory costs. Competitive pricing will usually depend on having a second source. Bringing up a second source may not be cost-effective unless the product is used in high volume or significant cost savings are expected through competitive bidding.

Figure 9-2 illustrates a relative unit pricing versus production volume curve representative of what might be encountered. For discussion, assume there are two vendors able to supply a particular custom IC and that the volume pricing of both vendors follows curve A in Figure 9-2. For simplicity, it will be further assumed that there are no NRE charges to bring production up at the second source (usually not the case). If the total production volume in question is 10,000 units per

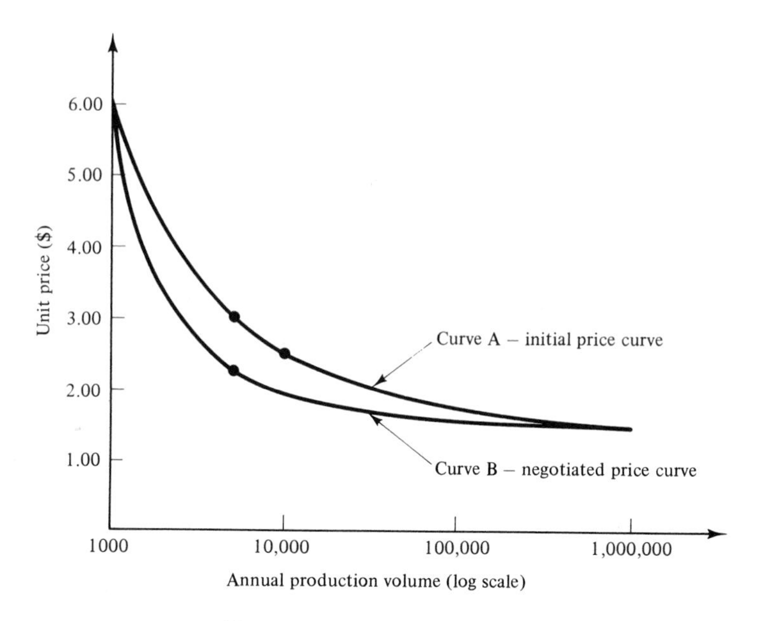

Figure 9-2 Price versus volume.

year, the IC cost would be approximately $2.50 from either vendor. If half the units were ordered from each vendor, the volume at each vendor would be reduced to 5000 per year and the price would increase to $3.00 per unit. The security of two vendors would, in this case, cost $0.50 per unit.

It may be possible to reduce this cost or actually achieve a savings through competitive bidding. One or both vendors may modify their pricing as shown in Figure 9-2, curve B, in an attempt to attract or keep the business. In this hypothetical scenario the units would cost $2.25 each, which represents a net savings of $0.25 per unit over the original price.

The process of defining an ASIC is one of maximizing the economic opportunities while minimizing the risk. Many technical risks can be mitigated by carefully partitioning the electronic portion of a product (system). There are many ways to approach the partitioning and definition of the portion of a system that is a candidate for integration. A "quick and dirty" approach might be to circle the circuitry that appears integrable on a copy of the existing schematic. This approach can give an initial guess but leaves much to chance. Many times, valuable alternative approaches are neglected because of preconceptions about a particular architecture or technology.

A more objective approach is to start by defining a detailed description of system requirements. Once the overall system-level requirements and functions are defined, it is relatively easy to allocate requirements to subblocks of the system. Next, break the top-level function into major subfunctions, such as power supply, amplifier, tone generator, and sensor amplifier. Break the subfunctions into building blocks, such as 5-V regulator, 12-V regulator, preamp, power-amp, and levy the appropriate requirements.

Consider the feasibility of meeting the allocated requirements and the required circuitry. Can the circuitry be partitioned to take advantage of standard ICs and still meet the space and power requirements? Do not reinvent the wheel if it is not necessary. Understand the requirements that are driving the integration effort (size, power, protection, performance, cost, market appeal, the competition, etc.). Be careful not to overdesign or overspecify the building blocks. There should be adequate margin to ensure necessary performance but not so much margin that the design becomes unnecessarily complex and expensive. The goal is not to design the perfect product. Make the product technically adequate and economically high performance, not the other way around. Finally, consider manufacturability, reliability, and future options or expansions.

A carefully planned and defined IC will have the best chance of being successful. The chance for technical success will be the greatest when the requirements of the chip are identified in advance. The most appropriate technology, design approach, and vendor can then be chosen. Unrealistic expectations will be identified early in the development cycle instead of at the end. More accurate bids will be received from the vendors. This will make the financial analysis more accurate. Careful planning will allow the customer to maintain control of the development effort.

9.2 Estimating the Value of a Custom IC

The value of using a custom IC in a product depends heavily on the product. For example, replacing a 3-in.-square circuit board with a single ASIC in a surface-mount package would have much more impact on the market success of a hearing aid than a refrigerator. Although significant cost savings might be achieved in either case, the hearing aid would not be a viable product without the custom IC.

There are several ways by which a custom IC can add value to a product. Some are rather obvious and their value added easily quantified. Others are less obvious and their contribution to product value and therefore the value of the custom IC is highly subjective. It would be difficult to list all potential value adders. The ones listed below are usually most significant. Keep in mind that the total value (in dollars) of any adder or cost saver depends on the size of the product's market window.

Once the potential custom IC has been defined, understanding the value it has is the next step. This information plays a major role in determining the wisdom of proceeding with the custom IC development. Knowing the value a custom IC has in a product allows vendor quotations to be accurately assessed. There is a solid criterion against which to make decisions. A thorough value analysis also helps to expose false assumptions about the merits of a custom IC. Analyzing the IC's value in a product has the further advantage of helping to optimize the definition of the IC, thus maximizing its value.

The total benefit of a custom IC is the combination of cost savings (if any) and value added to the product. Cost savings are fairly straightforward to calculate. Accurate estimates can usually be made if exact figures are not available. It is important to be sure that all applicable costs are included in the analysis. Estimating the value added to a product is a significantly more subjective process. Care must be taken to give an honest appraisal.

The following are examples of potential cost savings that might be realized with a custom IC.

- **Parts cost** Many people simply add up the cost of the parts replaced by a custom IC to determine the parts savings. This is a good start but is only part of the story. There are other significant costs associated with parts than just their purchase price. Many companies perform an incoming inspection on parts. The people, test equipment, and floor space dedicated to this operation add a significant cost to the parts. The floor space, recordkeeping, and finance charges associated with maintaining a large volume of many different piece parts also add cost. The variations of lead times for delivery of piece parts can have a significant impact on cost. Concern over shutting down an assembly line due to an unpredicted glitch in parts deliveries typically pushes inventory levels up. Fewer parts could mean smaller, simpler, and less expensive PC boards. This may not be the case if the product's package or PC board mounting is fixed. The board, although simpler, would be the same size.

• **Assembly cost** Smaller or simpler PC boards travel down the production line much faster than larger, more complicated ones. There are fewer parts to kit prior to inserting them in the board. Likewise, there are fewer parts to insert in the board prior to soldering. Production yields will be higher since there are fewer solder connections and fewer components. This means less troubleshooting and rework on the manufacturing floor. When rework is necessary, it is much easier to troubleshoot a board with fewer parts. Board testing will be simplified since a large portion of the circuitry, the custom IC, comes pretested from the vendor to the customer's specifications.

• **Product improvement** Products containing a significant amount of electronics can be improved in many ways with a higher level of integration. Consolidating and reducing the size of the electronics can allow the mechanical envelope of the product to be reduced. This may be of significant value, as in the case of the hearing aid, or of minimal value, as in the case of the refrigerator. Be sure to include any tooling costs (injection molds, etc.) in the overall cost calculations if a change in the physical size of a product is planned. A custom IC might allow a significant reduction in the required operating voltage or power consumption of a product. This could reduce the cost of power supply components. In the case of battery-operated equipment, the number and size of batteries might be reduced and/or battery life increased. This could also reduce the size and weight of the product, giving it more market appeal. Increased capabilities such as added features may increase the market appeal of a product and make it easier to use. A custom IC is inherently more rugged than a board full of discrete electronics. A product with a significant amount of its electronics integrated will be more reliable since there are fewer components and fewer mechanical connections. For the same reasons, a product with a higher level of integration will be easier to maintain. There are fewer parts to troubleshoot, repair, or adjust.

• **Future growth** A well-thought-out custom IC can serve as the basic building block for many products if the customer has a line of similar or evolving products. The basic architecture of the IC can be defined so that a portion of it can be used for simpler, less sophisticated products while the entire chip and additional circuitry (or an additional custom IC) can be used in more complex products. The chip could also be designed to adapt to anticipated future product changes.

These considerations may require that a few components that could otherwise be integrated be left external to the chip, or that a slightly larger package with more pins be used. Be sure to include the increased useful life of the IC in the cost trade-offs. An adaptable custom IC may be a little more expensive initially but may save many tens of thousands of dollars in redesigning and reintegrating the IC to adapt it to changing needs. The production volume of the more universal IC will be higher than the more specific IC and its unit cost will tend to be lower.

Total Potential Value

The following section gives an example of a value estimation for a hypothetical analog semicustom array to be retrofitted into an existing product. This example is not meant to include all possible considerations but should be representative of a typical analysis. The results of this analysis will be used again in Section 9.4.

A medium-sized company has a product made with discrete electronic circuitry in production with a run rate of 25,000 units per year. The product has achieved good market acceptance and is expected to continue selling at its current rate for approximately two more years. The lead engineer assigned to this product suggested that a significant cost savings might be possible if a large portion of the product's circuitry were integrated into an analog ASIC. The company's chief engineer asked him to investigate the potential cost savings and to quantify them.

This analysis identified the following per board cost savings

Parts cost

Total purchase price of parts integrated	$4.80
Incoming inspection costs (per board)	
(#parts tested) × (test time/part) × (labor rate)	$1.00
cost of reclaimable test floor space	$0.10
cost of scrapping/returning reject material	$0.10
Inventory costs of parts integrated (per board)	$0.15
interest costs on inventory	
cost of reclaimable inventory floor space	$0.10
cost of tracking and ordering inventory	$0.25

Assembly cost (per board)

Part kitting	$0.10
(#fewer parts) × (time/part) × (labor rate)	
Board stuffing	$1.17
(#fewer parts) × (time/part) × (labor rate)	

Estimated savings for board soldering	$0.58
(#fewer parts) × (time/part) × (labor rate)	
Estimated savings for board testing/alignment	$0.33
(time reduction/board) × (Labor rate)	
test equipment savings	$0.05
Estimated savings in rework	$0.03
(time reduction/board) × (labor rate)	
savings in scrap amortized per board	$0.01
Value added	
Miscellaneous cost savings	
reduced battery cost	$0.00
reduced power supply cost	$0.10
Estimated reduction in maintenance/warranty work	$0.10
(time reduction) × (labor rate)	
(administrative time) × (labor rate)	$0.05
postage and handling	$0.05
Estimated value added to customer (higher sales price)	$1.00
Estimated value added due to increased security	$0.10
Total estimated value/unit of product due to custom IC	$10.17

This total represents an estimate of the value the custom IC will have on a per unit of product basis (assuming one custom IC per product). The unit production price paid for the custom IC, including all amortized development costs, must be less than this figure to ensure that the custom IC is cost-effective.

9.3 Estimating the Cost of a Custom IC

The cost of the custom IC can be divided into two major categories: nonrecurring engineering costs (NRE) and recurring costs. The NRE is comprised of the costs incurred by the ASIC user to investigate, develop, and put into production a custom IC. Some of these costs will be expended within the ASIC user's own organization. Other costs will be paid out to the ASIC vendor for prototype development. The amount paid to the ASIC vendor will depend on whether the vendor does the circuit design as well as the prototype integration. The choice of whether or not to have the ASIC vendor or an outside design shop do the circuit design depends on many factors and will be discussed in more detail later. The recurring costs are the normal production costs to the ASIC user. They include costs such as the unit price of the ASIC and costs associated with any normal receiving inspection or test procedures.

All of the costs associated with a custom IC must be understood if an informed decision about the cost effectiveness of a custom IC is to be made. Many people look only at the prototyping charge and production volume price quoted by the vendor when they consider costs. There are, however, a large collection of "hidden costs" that must be considered to have an accurate picture.

The old adage "time is money" directly applies to custom ICs. Remember that any time spent thinking about the custom IC, requesting literature from vendors, meeting with the vendor or sales representatives, or defining the IC costs money. If an ASIC vendor or sales representative gives a "free" one-hour technical or sales seminar to five engineers, the real cost of this meeting is five fully loaded man-hours. When anyone in the organization starts looking into or considering a custom IC, the meter is running. Whether the circuit design is done by the customer or the vendor, there will be significant costs associated with the development effort. Some of the development tasks can be contracted to the vendor or an independent designer, others must be done by the ASIC user. The NRE costs begin to accumulate when the first investigation into the feasibility of using a custom IC starts and concludes when the idea of a custom IC is abandoned or when the IC is in production. This process can be broken down into five steps through which the development must progress.

- **Investigation** The investigation is the first step of the development process. During this phase, feasibility studies, vendor research, preliminary IC definition, specifications, and preliminary cost analysis are done. This step is very important because it lays the groundwork for future steps and determines whether or not the development of a custom IC is warranted. A thorough job at this point can save a significant amount of time and money later. At each successive step the cost to abandon the development gets higher.
- **Vendor selection** This step includes preliminary vendor selection, soliciting quotations, analyzing the quotations, negotiating with vendors, and making the final vendor selection. This step will involve some travel for vendor facility visits and several meetings with the vendor and his sales representatives.
- **IC development** The IC development phase includes finalizing the custom IC definition, specifications, packaging, test require-

CHAPTER 10

Choosing an ASIC Vendor

10.0 Introduction

The ASIC vendor that is ultimately selected will have a major impact on the success of the custom IC. The selection must be based on both the business and technical requirements of the custom IC development. The most elegant technical solution is worthless if it is too expensive or the vendor is uncooperative. Likewise, the most economically attractive option is unacceptable if it is technically unsatisfactory or if part delivery is unpredictable. The custom IC development process is a partnership between the vendor and the customer. This is especially true when the ASIC vendor is doing the IC circuit design. The performance of the IC depends both on the IC process and circuit design and the partitioning of the system (product) in which the IC resides. Trade-offs between product partitioning and IC specifications will be necessary. These can best be made by mutual agreement considering the impacts in both areas.

The choice of ASIC vendor should not be made at the last minute after the design is complete (as a PC board vendor might be chosen). The selection process extends from the IC definition phase to the actual order placement. This is true whether the vendor or the customer does the actual circuit design. If the customer elects to do the circuit design, a vendor should be selected prior to the start of design work. The vendor can provide a large amount of guidance during the circuit design phase. The vendor's engineers undoubtedly will have seen (and made) a great many design errors over time and can share this wisdom. Informal design assistance via phone calls or "sales" meetings is generally free of charge and can be very valuable.

10.1 Initial Screening

The vendor selection process begins during the early part of the ASIC definition stage. Many vendors would be delighted to help with the IC definition as part of their sales effort. The initial screening process should follow these basic steps.

- **Identify candidate vendors** Four to six potential vendors should be identified at the start of the IC development process. Vendor names can be obtained from this book, trade publications, advertisements, referrals, or previous experience. Try to identify all potentially viable vendors based on product and service offerings.
- **Collect literature** Contact the perspective vendors and ask for technical and sales literature. Make this effort the start of a vendor survey. Start a log. Keep track of how long it takes to receive literature. Was the literature what you requested? Do the people you talk to seem helpful and knowledgeable about their products? Review the literature you receive. Does it contain enough technical detail? Does it give information about the company, its history and future plans? Do you need more information or did you get what you need? Write all of your questions and comments in your log. Some companies may want you to purchase detailed technical manuals. These are usually inexpensive and contain sample transistors and macrocells you can evaluate. If a vendor you are strongly considering offers this detailed technical material for sale, you should seriously consider ordering it.
- **Get answers to technical questions** Make a list of technical questions. These questions may come from the IC definition activity or as a result of reviewing the technical literature. Put the questions and the vendor responses in the log book.
- **Get answers to economic questions** At this point you may be ready to send your preliminary IC definition and specifications to the vendors for a proposal and a quotation. This will serve two very important functions. First, the vendor responses will serve to validate the IC definition. You will have looked at it from a product/system perspective. They will look at it from an integrated circuit point of view. Any major definition errors should be caught at this point. The vendors

may make suggestions, that you had not considered, to make the custom IC more cost-effective. Second, you will receive a preliminary quotation for development costs and unit price that will allow you to complete your cost analysis and make the decision whether or not a custom IC should be pursued. Be sure to obtain information about prototype and production delivery schedules. Be sure you understand the quotation. Make sure it is clear what is included and what is not included. This information will be critical when you are comparing vendor quotes.

At this point, the initial list of four to six vendors should be trimmed to one or two. If the custom IC appears to be technically feasible and economically viable, you are ready to proceed with the final vendor selection.

10.2 Questions to Ask Potential Vendors

It is important to gather data about the vendors or vendor under consideration for the analog ASIC development. Information about the vendor's history, experience, financial security, future business plans, and business practices will give the potential customer a valuable insight into how the vendor will be to work with. Even if there is only one acceptable vendor, this information will allow the customer to anticipate, and plan for, potential problems that might occur during the development process.

Background Questions

- How long has the company been in the analog ASIC business?
- How many ASIC developments has the vendor completed?
- Will the vendor supply a list of past customers?
- Will the vendor give customer references?

Administrative Questions

- Is it easy to contact the right people (sales, engineering, etc)?
- Do you have a single point of contact within the vendor?

Technical Questions

- Does the vendor understand your requirements?
- Does the vendor have an adequate technology?
- Does the vendor have adequate technical information available?
- Does the vendor have experience with similar types of projects?
- Does the vendor process its own wafers or purchase them?
- Does the vendor have multiple wafer sources?
- Does the vendor package its own parts or subcontract assembly?

Economic Questions

- Is the vendor's business good?
- Does the vendor have a large backlog?
- Has the vendor recently introduced new products or services?
- Does the vendor advertise in major trade publications?

10.3 Vendor Facility Visit

The purpose of the visit is to get an overall feel for the potential vendor and to meet the people you will be working with. The comments in this section apply to the IC vendor and to an independent design service. Companies to which you subcontract work are in essence business partners with you. The success of your ASIC development depends on their professionalism, competence, and desire to meet their commitments. A face-to-face meeting is one of the best ways to ascertain their ability to provide the help you need.

- **Meet the people you will be working with** If the vendor will do the circuit design, ask for resumes of the design engineers who will actually be doing the design work.
- **Arrange a technical meeting to discuss the proposed IC** This will give you an opportunity to work directly with the vendor's engineers on a technical level. This meeting can be considered not only a working session but also a technical interview of the vendor's engineers. You can get an understanding of their approach to your design and their level of competence. This is a very important step if the vendor is doing the actual circuit design.

- **Arrange a tour of the vendor's facilities** Ask to see the design lab. Look for the test equipment in use. Check to see if it has been calibrated recently. Is it modern and in good repair? Is there enough equipment or does it appear that a lot of the equipment is shared? A shortage of equipment can mean delays in design work. Look at their CAD tools and CAD facility. Are simulation tools extensively used? Are DRCs always run on a layout? Ask to see their automatic test equipment (ATE) facility. Are ESD (electrostatic discharge) safety procedures being followed? If you are unable to go into the wafer fab facility, can you look through windows? Is the equipment clean and in good repair? Are the employees busy? Do they appear to be efficient? Are there written work directions at each work station?

General Observations

After the tour is over, summarize your observations in your log. Were you treated courteously? Were your questions answered completely? Were the work areas organized, neat, and clean? Did the employees appear busy and happy? Did people seem harried and stressed? Did meetings start on time? Were meetings organized and did the appropriate people attend?

10.4 Negotiating

There are four areas that must be negotiated: NRE, production pricing, prototype delivery, and production delivery. The starting position will be that the vendor wants the highest prices, the largest volume commitment, and the most flexible delivery schedule possible. The customer, on the other hand, wants the lowest possible prices and the shortest possible deliveries. Somewhere between these extremes is a position that will be acceptable to both parties.

Negotiate NRE payments in stages. Make an initial payment of 25 to 50% when the initial order is placed. Then make progress payments at major milestones such as circuit design approval, IC layout complete, functional prototype delivery, and final test sign-off. This allows the customer to maintain some control over the development process and gives an incentive to the vendor to complete the integration quickly.

Prototype and production delivery will generally depend on the vendor's manufacturing cycle and is not overly flexible. Deviating from

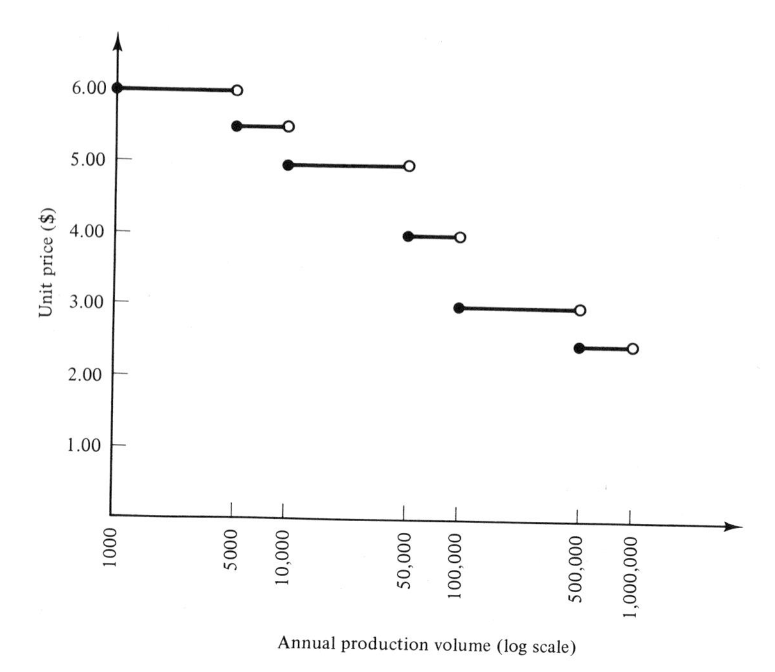

Figure 10-1 Hypothetical price versus volume quote.

the normal scheduled production flow with "hot lots" to reduce manufacturing time is possible but can cause problems for the vendor. These special runs have increased risks of processing errors and could, as a result, end up in the scrap bin.

The vendor will want a large production commitment and will offer price incentives to encourage larger orders. These incentives can include reductions in design engineering or prototype integration costs and/or lower unit prices for large quantity orders. If you accept lower NRE costs, be careful that you do not pay an inflated unit price. This could be far more expensive in the end.

There are several strategies a vendor might use when negotiating volume production pricing. A falling price versus annual volume, as illustrated in Figure 10-1, may simply be quoted. In this case, the vendor will expect the agreed-upon volume to be scheduled for delivery over a 12-month period. If the deliveries are later reduced, the vendor will want to renegotiate the unit price upward. If, on the other hand, the quantities are larger than originally anticipated, the customer will want a price reduction. The vendor may propose a "bill-back" clause in the volume agreement. This would allow the vendor to bill the customer for the difference in unit price if some negotiated minimum quantity was not purchased. For example, the initial agreement was to purchase 10,000 parts at $2.50 each. The vendor also quoted the 5000 unit price at $3.00. Under a "bill-back" arrangement, if the customer ordered 7000 parts instead of 10,000, the vendor would bill the customer the difference in the unit price times the number of units actually ordered. In this example, the customer would receive a bill for $3500. Some times the vendor will renegotiate production orders in lieu of a bill back. Another strategy is called "step pricing." With this pricing technique, the customer pays one price for the first x pieces, a lower price for the next y pieces, and a still lower price for the next z pieces, and so on. The prices are such that the average price over the agreed-upon volume is the agreed unit price. This pricing technique protects the vendor from customers that negotiate a 100,000-piece price and then actually take delivery on 10,000 parts. The disadvantage to the customer is the up front high cost and the need to carefully monitor the pricing to ensure that the billing is correct.

It is wise, when negotiating the production order, to allow an "out" in case the new product does not sell as expected or there is some kind of business reversal. You do not want to get stuck with 100,000 custom ICs you cannot use.

Final Vendor Selection

Two decisions remain after price and delivery negotiations have concluded. Are the price, delivery schedule, and risks consistent with the anticipated value of the IC and the market window for the product? In other words, is developing the proposed custom IC a desirable option? Second, if you decide to proceed, which vendor will you select? The results of the value assessment, technical evaluations, and information maintained in your log should provide a sound basis on which to make these decisions.

10.5 Managing the Development

Careful management of the development is critical, once the decision to proceed has been made and a vendor selected. There are many opportunities for misfortune during the circuit design and integration even if the IC and its associated requirements have been carefully defined and the vendor and design resource chosen with great care. The price of success in an ASIC development is unrelenting vigilance. Open communications with the design resource (whether it is the vendor, an independent resource, or an in-house design team) is extremely important. The development effort should be headed by a single responsible person. During the design phase, there will be opportunities to make trade-offs between chip requirements and system requirements. Decisions in these matters must be carefully considered lest a costly oversight creep into the design. The later in the design any change is made, the more chance for error. A well-defined approval procedure must be established at the start of the design effort. Any and all design changes and the logic behind them should be carefully documented for future reference. This is best accomplished through a committee made up of the designers actually designing the IC, the system engineering group responsible for the end product, the manufacturing organization, purchasing, and the marketing or sales department. Any change may affect some or all of the groups.

Plan the development in stages with tangible results at the completion of each stage. You should have something in your hand that can be approved at the completion of a milestone. The stages or milestones should be serially dependent. A given step cannot be started until the previous one has been satisfactorily completed. Design efforts with several activities such as product definition, test development, circuit design, and market studies occurring in parallel can easily become out of control, especially if some of the effort is done through subcontractors.

Carefully planned milestones serve two purposes. The first is to measure the successful progression of the design. At each milestone, all in the approval loop can review the development and approve its progress from their particular perspectives. The technical progress and logical development of the design will be well documented. Second, the development progress with respect to time can be measured. This is critical if a significant savings can be realized by putting the integrated version of the product into production or if there is a limited market window for the product.

Some suggested milestones are as follows:

Milestone	Measure
Initial feasibility determined	Feasibility study
Initial IC definition complete	Block diagram/specs
IC value determined	Value analysis
IC cost determined	Cost analysis

Decision to develop the IC made	Analysis complete
Vendor selected	Order placed
Design resource selected	Order/contract
Development committee selected	Members chosen
IC definition/specifications finalized	Documentation
Preliminary circuit design complete	Schematic
Simulations complete	Analysis results
Worst-case analysis complete	Analysis results
Final circuit design complete	Schematic/specs
Test plan complete	Test plan
Layout complete	Layout
Product update complete	Design documents
Prototype ICs delivered	Devices received
Prototypes characterized	Characterization data
Tests and specifications finalized	Test definition/specs
Initial production deliveries scheduled	Production contract

Index